The Kentish's of Keysbrook

A history of the Kentish family who settled in the Keysbrook district of Western Australia.

By
David Kentish

Preface

The writing of a book recording the history of The Kentish's of Keysbrook has been in my mind for some years and since we moved to Forrestfield in November of 1994, when we leased out our Williams piggery, I was able to put some real-time to the project.

Unfortunately, both of my parents, Lance and Vera, had passed away a few years before me being able to begin.

The main concept of the book is a recording of the family history from the memories of the surviving family members. With Dad and Mum gone, I was already at a disadvantage but was determined to complete this project.

During the reading of the book, you will notice that I have not always been specific with exact dates of some events. This is because I have worked from interviews with Enid, Clem, Esther and others and exact dates are not always in memory. Most dates are complete and these are readily confirmed by researching other records and printed material.

To fill in some blanks where the memories of Mum and Dad are concerned, I have referred to Mum's diaries, which she commenced in 1961. Gwen (dec.), Marjory, Lorna, Neil and I have spent a few hours going through old photographs and notes to fill in these gaps and I believe that I have been able to present a fairly accurate picture of the life and times of The Kentish's of Keysbrook.

Part one of the book relates to dates up to 2009 and this edition brings things up to date, except the Family Tree, which I have elected to leave as it was in 2009

Mostly I have attempted to show you the daily lives and experiences of the family. These are the types of records, which can't be gleaned from history books.

The compilation of the details and the writing of this book would not have been possible without the help of my wife, Barbara, whom I thank very much for her help in assisting with the preparation and proofreading of the material and her understanding when I didn't answer her because I was engrossed in my task.

While I have taken good care to be accurate with all details, I do apologise if I have missed something or have recorded it incorrectly. Even though I may have taken some literary license with a few details, the stories are correct.

I would like to dedicate this book to the memories of those Kentish's of Keysbrook who have given us so much and passed before us.

Edward Joel Kentish, my grandfather; Edith Alice Kentish, my grandmother; James Lancelot Kentish,

my father; Katherine Vera Kentish, my mother; Ira
Eleanor Elizabeth Kentish, my aunt; Reg Ingpen, my
uncle.

And since 2006, Herbert Clement Kentish, my uncle;
Stuart James Kentish, my nephew; Gwendoline
Alice Swaby, my sister.

David J Kentish
P O Box 359
Goomalling 6460
Western Australia
dkentish@westnet.com.au
www.davidkentish.com.au

Coat of Arms

There are a few variations of the Kentish Coat of Arms and it is not clear as to whether we are entitled to have the use of them or not but they have been used for many generations on Crests, Monuments, Family Plates and Bookplates. A description of our Coat of Arms is "Gules, a pair of wings conjoined in lure ardent overall a bendlet azure. The crest being demi ostrich wings endorsed, holding a horseshoe in its beak"

The Kentish Family Tree

A completed Family Tree has been compiled and reported on by Mr Peter Kentish in his book "Passage for Prosperity". For this history book, I will start with James Kentish, First child of Thomas Hirons and Mary Clay Kentish and continue along this family lineage.

1st Child of Thomas Hirons and Mary Clay KENTISH

James Kentish	Married	Elizabeth Barrett
Born 04/08/1801	1822	Born 21/01/1800
Died 23/12/1878		Died 08/08/1885

Issue: James (Junior), **David Joel**, Elizabeth, Charles, Henry, and Emma Mary

---

2nd Child of James and Elizabeth KENTISH

David Joel Kentish	Married	Anne Maria
Born 16/10/1824	07/08/1851	Horwood
Died 31/01/1903		Born 1827
		Died 06/03/1879

Issue: Henry Horwood, Agnes Clay, Annie Horwood, Evangeline Horwood, Sarah Helen, Frank Butler, Grace Horwood, Herbert David, **Edward Joel**, Alice Martha, Lydia Horwood, Ernest Horwood.

9th Child of David Joel and Anne Maria KENTISH

Edward Joel	Married	Edith Alice
Kentish	20/02/1908	Nicholls
Born 20/03/1865		Born 25/03/1877
Died 26/07/1942		Died 26/01/1960

Issue: Enid Winifred, James Lancelot, Herbert Clement, and Esther Bell.

1st Child of Edward Joel and Edith Alice KENTISH

Enid Winifred Kentish Born 28/07/1910	Married 04/11/1944	Reginald Robert Courtney Ingpen Born 24/12/1901 Died 11/01/1996

Issue: Roger Edward, Joan Winifred, Edward John, and Robert Frederick.

Roger Edward Ingpen   Born 11/10/1945 died 16/10/1945

Joan Winifred Ingpen Born 13/03/1947	Married 06/10/1973	Peter George Spooner Born 21/06/1942

Issue: Heidi Michelle, Alexander John, and Reginald George

Heidi Michelle   Born 18/02/1975
Alexander John   Born 19/09/1978
Reginald George   Born 22/07/1982

Edward John Ingpen Born 23/08/1948	Married 13/03/1980	Sandra Bowman Born 09/11/1950

Issue: Nicole Lee, Kylie Bowman, and Benjamin Bowman

Nicole Lee   Born 01/03/1982
Kylie Bowman   Born 17/03/1975 (Adopted)
Benjamin Bowman   born 17/03/1977 (Adopted)

Robert Frederick Ingpen Born 07/01/1952	Married 11/04/1992	Margaret Ruth Clarke Born 17/05/1948

2nd Child of Edward Joel and Edith Alice KENTISH

James Lancelot Kentish Born 25/07/1912 Died 23/09/1990	Married 13/03/1937	Katherine Vera Lavis Born 27/08/1912 Died 21/03/1991

Issue: Gwendoline Alice, Marjory Ruth, Lorna Joyce, David Joel, and Neil James

Gwendoline Alice Kentish Born 05/04/1938	Married 01/06/1957	Frederick William Swaby Born 29/09/1928

Issue: Audrey Lynette, Eric John, Doreen Rose, and James William.

Audrey Lynette Swaby Born 31/05/1958	Married 24/11/1984	Linton James Tuckey Born 02/12/1960

Issue: Matthew Robert, Amanda Deborah, and Rebecca Jane

Matthew Robert   Born 15/07/1988
Amanda Deborah   Born 07/06/1990
Rebecca Jane   Born 27/03/1995

Eric John Swaby Born 12/04/1960	Married 15/11/1981	Susan Maria Hernandez Born 30/06/1962

Issue: Jaylene Brooke born 25/03/1992

Doreen Rose Swaby Born 16/07/1961	Married 12/01/1980	Gary Donald Fieldstaff Hughes Born 08/07/1958

Issue: Craig Garry Fieldstaff, Kathy Margaret

Craig Garry Fieldstaff   Born 06/05/1980
Kathy Margaret   Born 22/09/1983

James William Swaby born 29/06/1974

Marjory Ruth Kentish Born 26/04/1941	Married 1 21/09/1963	Gregory Clarke O'Neill Born 21/06/1939
	Married 2 02/02/1977	Kenneth Peters * Born 03/04/1932
	Married 3 22/09/1991	Ronald Tree * Born 04/09/1924

Issue: Stephen Bruce, Rodney Ian, Craig Anthony, Wayne Gregory
* No Issue

Stephen Bruce O'Neill Born 23/03/1964	Married 1 22/11/1986	Sheryl Shaddick Born 06/09/1958
Issue: Clinton Born 17/12/1982 (Adopted)	Joshua Born 28/02/1986	Nathan Born 21/10/1987
	De facto	Anne Marie Byrne Born 07/01/1967
Issue: Lance Ryan Born 14/11/1992	Thomas Christopher Born 11/08/1994	

Rodney Ian O'Neill Born 26/09/1965	Married 10/11/1989	Jacky Boladeras Born 21/08/1965

Issue: Kristie Lee and Sally Anne both born 23/01/1991 (Twins)

Craig Anthony Born 28/09/1967
Wayne Gregory Born 20/06/1969

Lorna Joyce Kentish Born 12/07/1944	Married 26/10/1963	Albert Barry Robins 30/06/1938

Issue: Susan Deborah, Diane Sharon.

Susan Deborah Born 14/10/1964	Married 12/10/1985	Peter John Heal Born 10/10/1957

Diane Sharon Born 29/12/1968	Married 29/10/1989	Glen William Turner Born 26/10/1957

Issue: Liam Russell born 18/11/95

David Joel Kentish Born 27/01/1949	Married 19/10/1968	Barbara Rose McKay Born 09/04/1946

Issue: Kevin Ian, Bruce Mark, Vicki Leanne

Kevin Ian Kentish Born 05/04/1969	Married 03/03/1990	Alison Mary Hyatt Born 20/11/1960

Issue: Riley Blake   Born 02/08/1990
Hannah Ashleigh   Born 02/10/1992

Bruce Mark   Born 30/10/1972
Vicki Leanne   Born 03/02/1977

Neil James Kentish Born 18/11/1951	Married 23/11/1980	Jacqueline Saunders Born 18/01/1951

Issue: Josephine,   Born 25/03/1983
Stuart James   Born 14/12/1984

3rd Child of Edward Joel and Edith Alice KENTISH

Herbert Clement Kentish Born 22/05/1914	Married 01/05/1952	Ira Eleanor Elizabeth Forbes Smith Born 02/02/1920

Issue: Coralie Elizabeth, Colin Clement, and Heather Gaye.

Coralie Elizabeth Kentish Born 01/07/1953	Married 07/07/1973	Charles Graham Parkin Born 13/03/1946

Issue: Deborah Fluer, Michael Graham
Deborah Fluer   Born 25/08/1974
Michael Graham   Born 28/10/1978

Colin Clement   born 09/05/1955

Heather Gaye Kentish Born 07/03/1959	Married 06/09/1986	Colin Edward Payne Born 29/06/1954

Issue: Laura Ashlee, Emma Nicole
Laura Ashlee   Born 06/04/1988
Emma Nicole   Born 06/02/1991

4th Child of Edward Joel and Edith Alice KENTISH

Esther Bell Kentish Born 01/02/1917	Married 08/06/1940	Allan Cole Uren Born 15/09/1914

Issue: Beverly Joy, Lyal Edward, Allan Graeme, Lynette Grace, Shirley Faye, Glenda Belle, Valerie Merle, Kevin Ross.

Beverley Joy Uren Born 27/12/1943	Married 14/01/1967	Trevor Alexander Moffat Born 12/09/1943

Issue: David Allan, Danielle Gaye, Suzanne Joy, Andrea Bell, and Tracy Anne.

David Allan Moffat Born 11/11/1968	Married 09/02/1991	Pam Ryall Born 17/02/1969

Issue: Alycia Rose   Born 08/08/1993

Danielle Gaye   Born 30/12/1969
Suzanne Joy   Born 24/02/1972
Andrea Bell   Born 14/09/1975
Tracy Anne   Born 14/09/1975

Lyal Edward Uren   born 17/05/1945

Allan Graeme Uren Born 04/06/1946	Married 26/01/1971	Mary Patricia Keenan Born 12/10/1946

Issue: James Brendon, Brian Cole, Harleigh Graeme, Patricia Maree, and Damien Francis
James Brendon   Born 05/04/1972
Brian Cole   Born 16/05/1974
Harleigh Graeme   Born 23/07/1979
Patricia Maree   Born 30/06/1981 (Stillborn)
Damien Francis   Born 21/09/1983

Lynette Grace Uren Born 17/01/1948	Married 17/07/1971	David Albert Crozier Born 09/04/1943

Issue: Samantha Lynette, Christine Anne, Rebecca Michelle, and Roger David
Samantha Lynette   Born 02/02/1972
Christine Anne   Born 21/07/1974
Rebecca Michelle   Born 27/08/1977

Roger David   Born 08/05/1981

Shirley Faye Uren Born 28/12/1952	Married 14/06/1975	Colin Frederick Trigg Born 15/06/1950

Issue: Carleen Heather, Bree Renae, And Cade Ramsey

Carleen Heather   Born 01/12/1978
Bree Renae   Born 05/05/1981
Cade Ramsey   Born 05/04/1988

Glenda Belle Uren Born 26/03/1954	Married 14/05/1977	James Allan Cain Born 11/10/1952

Issue: Simon Peter, James Michael, and Samuel John

Simon Peter   Born 11/06/1979
James Michael   Born 22/02/1981
Samuel John   Born 03/03/1984

Valerie Merle Uren Born 24/11/1957	Married 31/10/1981	Ian John Taylor Born 02/01/1959

Issue: Christopher John, Casey Nicole, and Lisa Louise

Christopher John   Born 03/12/1984
Casey Nicole   Born 24/10/1986
Lisa Louise   Born 16/01/1988

Kevin Ross Uren Born 05/11/1959	Married 03/02/1980	Tanya Margaret Leman Born 08/10/1961

Issue: Joshua David, Rachel Shannon

Joshua David   Born 02/01/1987
Rachel Shannon   Born 02/12/1993
Melisa Elsie   born 30/11/1978 (Fostered)
Ebony May Louise   Born 01/03/1991 (Fostered)

Nicholls Family Tree

3rd Child of Sarah Elizabeth Rhodes and Douglas BELL

Joseph Bell (the Pathfinder) Born 18/01/1809 Died 22/08/1879	Married 25/02/1835	Ann Patton Born 3/01/1810 Died 21/08/1879

Issue: Joseph, George, Mary Anne, James, Sarah Elizabeth, Emily, Thomas, Annie Martha, and Amy Rhodes

3rd Child of Ann Patton and Joseph BELL

Mary Ann Bell Born 27/02/1839 Died 18/08/1906	Married 21/09/1858	James Nicholls Born 1831 Died 21/06/1911

Issue: James, Annie Louisa, Julia, William, Elizabeth, Alfred Herbert, Kate Emily, Marion, Ada May, Edith Alice, Harry, and Grace

10th child of Mary Ann Bell and James NICHOLLS

Edith Alice Nicholls	Married 20/02/1908	Edward Joel Kentish

Refer to The Kentish Family Tree for the rest of the details.

It all begins. South Australia

England is where our heritage lies and this history begins in the
16th century. We have descended from a line of Kentish's who
shared the Christian name of Thomas. From the earliest records
that are available, it has been determined that The Kentish
lineage begins with a Robert Kentish who married a Sibill
Crossebie on the 23rd of April 1599 probably in the Church of
St. Stephens at St. Albans, Hertfordshire, England. Six
generations of Thomas followed with the last having been given
his mother's maiden name of Hirons as a second Christian
name.

These people were mostly farmers and had farms named
Tenements, Shenish Farm, Aldenham Farms & Millhouse Farm.
These were in the vicinity of St. Stephens Parish in
Hertfordshire. As families grew some left the land and began
life in other trades and businesses.

Thomas Hirons Kentish married Mary Clay on September 3,
1798, at St. Bride, Fleet Street London, him being listed as a
Draper of 42 Ludgate Hill, St. Martins, Ludgate and she of Fleet
Street London. This couple had seven children: James, Thomas,
John, Mary, Selina, and Elizabeth & Martha.

James Kentish, whilst listed as a grocer, became married in 1882
to Elizabeth Barrett, whose father was a draper. James and
Elizabeth had 6 children: James (Junior), David Joel, Elizabeth,
Charles, Henry and Emma Mary.

In the year of 1837 on December 15, James, accompanied by his
wife Elizabeth and their six children left England aboard the
Canton bound for Adelaide in South Australia. The Canton was
a 3 masted sailing ship, approximately 37 metres in length with

a load capacity of 500 tonnes. This voyage took 140 days and their arrival and subsequent disembarkation at Holdfast Bay was on 1st May 1838. Holdfast bay (now Glenelg) had no unloading or docking facilities so all cargo and passengers were unloaded with rowboats and barges until becoming grounded then they had to wade or be carried piggyback style the rest of the way to the beach. Their goods were stacked on the beach above the high water line and moved later.

Their first impression of their new country was somewhat disconcerting when compared to the comforts, which were left behind in England. With no docking facilities, limited roads and vehicles to cart themselves and their goods and animals to their selections. Sand-dunes and creeks had to be crossed to reach Adelaide and the landscape was entirely dissimilar to that to which they were accustomed. You can just imagine the family standing on a long sloping beach with some bedding, furniture, tubs of crockery and other items scattered around them looking at the ship being unloaded in all of its disorganisations. Rowing boats and barges plying to and fro with men wading whilst carrying women and children on their shoulders to keep them from becoming wet. Other men were lugging goods and furniture upon their shoulders whilst wading between the rowboats and the shoreline.

Total apparent chaos.

Rough bough shelters in Adelaide gave way to tents supplied by the Governor for their first accommodation. To reach this accommodation they were required to traverse the sandhills, cross several creeks and walk through this alien landscape. The wildlife was also something to behold. Imagine seeing Kangaroos for the first time. Imagine seeing and hearing Kookaburras for the first time. That raucous noise of theirs may have frightened them. I'll warrant they were intrigued with it. Parrots and other native birds were abundant in the vast groves of native trees that they had never seen before. Just imagine the bewildered feeling.

In that first year of their arrival in Adelaide, James purchased a one-acre section No 98 fronting Grenfell Street. He was shown in the records to be a Dairyman, so we assume that they were supplying milk and milk products to the residents. The ages of the children at the time of landing would have been approximately: James (Jnr) 15 yrs, David 14, Elizabeth 12, Charles 9, Henry 8, and Emma 3. The older children would be a great help in getting things going in their new land.

The land was leased, bought and sold whilst building a satisfactory income from the land by way of mixed farming. Value-adding is something we think of as a new idea but that was the norm in those days to maximise the return for your efforts. Not only sell the milk but also butter and cheese, not only sell grain but also the flour and bran etc., all go to help the income.

David, the second child of James & Elizabeth Kentish, being in his 14th year upon arrival in Adelaide, would have been a great help to the family as was his brother James. Instead of payment for his efforts, David was gifted 41 acres of land at Goodwood from his father. This was to be the beginnings of his land transactions and farming in his own right.

It was whilst on this property that he married Miss Anne Maria Horwood. Anne was the sister of Mr J H Horwood who was the founder of J H Horwood & Co. (then onto Horwood-Bagshaw, the machinery manufacturer). We believe that Anne Maria and her brother J H Horwood originated from the city of Oldham in

Lancashire and arrived in Adelaide a few years before she became married. The first 4 of their 12 children were born at this property with the other 8 being born while they were working the Farm at Munno Para. This was called Prospect Farm in the Gawler Hills and where they lived for 22 years. During this time David was involved with affairs of the local district, resulting in several roads being named after the family.

David used to hunt wild animals to supplement the family food supply. A large busy family would have a large appetite. He wasn't always quite at home in the bush however and occasionally became disoriented with directions. After capturing his prey and becoming lost he would "coo-ee " out. His large voice would carry for quite a distance and soon someone would come to him and guide him home. I think this also was a good opportunity to have some help to carry the prey.

When Edward Joel was just 14 days from his 14th birthday, on March the sixth, 1879 his mother, Anne Maria, passed away. She was buried in the Unley cemetery where her headstone and grave have been preserved.

This must have been a great blow to David and his family as it was only a few months after her passing that the Prospect farm was sold and the family moved northwards to take up land in the Booleroo - Willowie area.

Moving north.

This farm was sections 73 & 79 south, of The Hundred of Willowie and was named Oldham Farm, probably after the City of Oldham where Anne was born and spent her childhood. Edward Joel or Ted as he was known along, with his brothers and sisters would be helping with the farm work and Alice is known to have kept house for David (her father) for many years.

We know that he, Ted, did receive some formal education as he was enrolled at The Stanley Grammar School at Watervale (South of Clare) in the year of 1883. As a full boarder, the cost of education in those times was 13 Guineas (£13/13/0) per quarter being paid in advance. It is also noted that both William and James Nicholls attended this same college but in the years of 1881 and 1885 respectively.

Ted was a powerful man of stocky build with very large hands with a heart to match. It has been said, "You know when you shake the hand of Ted Kentish because yours disappears". He was a willing worker with no job too small or too big to tackle. Ted was working Oldham farm with his father and siblings at least until 25 March 1894 when, with his brother Herb, he leased sections 3 & 61 in the Hundred of Appila from their Uncle James and worked this farm for about 7 years. It may have been at about this time when Ted and Herb went mining for silver, probably to raise money to assist in the purchase of land. When the lease expired Herb renewed it by himself and Ted returned to Oldham farm to assist his father who was now at the age of 77. At the passing of his father on 31st March 1903, Ted was willed Oldham farm.

The house was built from local sandstone pebbles held in place with "pug", a mixture of clay and sand, and the walls were rendered with cement to give them a smooth finish. This provided a wall of very solid construction. These stone walls were approximately 400 mm thick and kept the house quite cool. Timber rafters were used to support the iron roof with a bull-nosed verandah at the front. The house was sited on a low hillock close to the south boundary of the farm with the front of the house facing Mount Remarkable. This position allowed the afternoon breeze to flow through the house, thus cooling it, much to the relief of everyone.

Beneath the floor of one of the rooms was built a large cellar where they kept stores and food. As there was no refrigeration in those days much of the meat was smoked or soaked in brine to preserve it. It would keep for many weeks in the cellar. In a large cellar like this, there would also be stored some seldom-used household items.

Rainwater tanks were a must around Booleroo due to the lack of suitable shallow well water for household use. Ted had several rainwater tanks around the house. These would hold the water that ran from the roof of the house whenever it rained. During Ted's occupation of Oldham Farm, they re-modelled the kitchen. First, they built a large underground tank and using the roof of the tank to become part of the floor of the kitchen. Local gravel and sand from the creek along with cement would be used to make concrete for this. A large dam would provide water for the sheep, cattle and horses. Sometimes during a drought, these tanks and even the dam would become dangerously low and water would need to be carted from the Government Well at Booleroo, which was equipped with a " whim " (refer to the Nicholls story for a description of this). A steel tank would be fitted onto the back of a wagon that was harnessed to a team of horses. Ted used horse teams exclusively but there was also bullock and camel teams doing this and other haulage work.

As was the normal practice in those days the sheds and outbuildings were constructed with bush poles. These would form the mainframe of the sides and the roof. There were various roofing materials used, including brush or straw. Either was normally fitted between two layers of wire netting to hold the roof material in place. Both of these methods were called "Thatch ".

Being in the northern parts of the agricultural area of South Australia had many benefits but it also had some disadvantages and one of the worst was the frequent drought. In 1914 there was a bad drought and many local farmers were given the job to lay a 4-inch (100 mm) pipeline to deliver water to the majority of the area known as The Hundred of Willowie. Not only did this provide some farmers with drought relief in the form of cash for their labours but also water for their stock and household use. There is an account of a storm at the time of trenching and overnight the trench that had been dug by hand the previous day became filled with dust and debris. Next morning many men with long faces were cleaning out the trench, probably for no pay.

For many generations, the Kentish family have been of the Christian faith and have actively participated in their local church. Ted was no exception. He was a local lay preacher at the Booleroo, Willowie and Merles circuits. He also had time to give to the church in the form of Sunday school teaching and it was here that he met a young lady by the name of Edith Nicholls with "beautiful brown eyes". Some years later after a courtship, they became married on 20th of February 1908 in the Booleroo Whim church. Ted would continue with his work for the district and the Church until he left for Western Australia. There he carried on with the same work as he had whilst at Booleroo.

The family grows

All four of Ted and Edith's children were born while they were living at Oldham Farm at Booleroo Whim.

They attended school at Booleroo Whim. To get to school they would ride in the horse-drawn buggy the 3 miles or ride their horses. Sometimes when the creek was flowing high, they had to leave the horse and buggy at a friend's place near the creek and walk over the swing bridge, which was erected for this purpose. This bridge has long since gone but you can still see the trees that it was attached to with chains and cables. Sometimes the boys would get to the middle of the bridge, grab the top cable in their hands and rock back and forth. This would set up a swinging motion and cause some distress to the girls, much to the amusement of the boys.

On school days Clem and Lance would harness Daisy, the pony, to the buggy and she would trot most of the 5 kilometres to school. A neighbours' child, Molly Foulis, travelled with them for a time until she was able to ride her horse.

Mrs White was the teacher for a time at Booleroo Whim School and Clem remembers an encounter she had with Neta Quick. For some reason, Neta upset Mrs White, who had a violent temper, and she dodged a book thrown by the teacher. The book continued along its flight path and collided with a few jars on a shelf at the back of the room. In these jars were frogs, scorpions, snakes and lizards preserved in methylated spirits. You can imagine the frustrated teacher and imagine the mess and smell these would make as they smashed onto the floor.

The teachers must have been a short-tempered group in those days as Clem remembers an encounter that he had with a Miss

Sargent. She upset Clem and he retaliated and ended by kicking her in the shins. She wrote a letter home to his father but he ran all the way home to tell his Dad his side of the story before the letter arrived. He doesn't recall any action taken by his Father.

As the School was only a hundred metres or so from the creek, this was deemed "Out of Bounds ". Not far from the schoolroom was a gravel pit, which left a depression in the ground. Clem and the other boys encouraged all the children to hide in this gravel pit. When the teacher came out to ring the bell to return to class, she could not find any children as they all laid low in the old pit. Assuming they had gone to the creek, she searched there for quite some time. Eventually, all the kids stood up and cheered as the angry, fuming teacher returned them to class.

On Sundays, the boys would harness two horses "Teddy and Dick" to a larger buggy and the whole family would head off to Church. The horse would pull the buggy at about 15 Kilometres per hour. Sometimes after church, they would go to Nicholls' for the afternoon and returning home in the evening. The children would attend Sunday School and they have fond memories of the Sunday School anniversaries. These would have been the highlight of the Sunday School year. They would learn to sing special hymns to sing to their parents and friends. The girls were fitted out with new dresses and the boys had new shirts especially for the occasion. Miss Bodkin was a local woman who made the clothes for the children of the district and Esther was frustrated at having to stand still for so long at a fitting to have the hem put up.

Strawberry Fetes were a feature of the church. Obviously to raise money for the operation of the church and church activities. These fetes were held at intervals to coincide with the availability of strawberries. They would be sent up by train, arriving mostly in good condition. As some farmers had milking cows, the cream would be abundant. What a feast!! Strawberries and cream!!

Enid sometimes rode Daisy but at the crossroads, she jibbed and stopped. Lance would then have to touch the whip to the horse's rump to get her going again. Enid received a new whip for Christmas one year and was very proud of this. She was required to assist in the house to help her mother and one of these jobs was to do the dishes after the evening meal. Enid did not like doing the dishes. Ted had become frustrated with Enid repeatedly not doing her job, so after a time he took her new whip and used it on her legs. This appeared to cure Enid of her problem as she then did the dishes without complaining.

Another of Enid's jobs was to wash the lamp glasses each Saturday. Kerosene was burnt in the lamps to provide light in the house. This burning kerosene gave off a faint black smoke, which discoloured the glass around the lamp and to give the best light, these needed to be kept clean. So this was Enid's Saturday morning job.

Ted being the very gentle man had a wild pigeon fly down & perch upon his shoulder. He didn't flinch but his puzzlement ceased as after a few minutes the pigeon flew off and when Ted looked up into the air he spied a hawk. Obviously, the pigeon was seeking refuge from the worst of its enemies.

Each of the family members was helping on the farm or in and around the house, so Ted would make sure that they all had an annual holiday. This usually was a trip to Glenelg where they had the use of a cottage near the beach. The cottage was easy to recognise as a light pole outside was painted in zigzag stripes. Clem & Lance convinced the girls that a can of striped paint was used for this purpose. Ted would drive the family to Booleroo Centre to travel to Adelaide on the train. Sometimes Ted would accompany them on their holiday.

It was whilst on one of these holidays that Esther and her mother caught a tram to Adelaide and were walking down the street when Esther saw in the window of a shop a doll. This was one of the finest china dolls that Esther had ever seen and it was love at first sight. They went into the shop and Esther purchased the doll for 10 shillings of her own money, a birthday gift. Her mother thought that this was very expensive but Esther was satisfied and that thought mollified her mother. The doll was with Esther for many years.

Swimming was very popular at Glenelg and because of the contrast to the land at Booleroo, the beach was very welcome.

Clem had an early brush with the Police as one day he became lost while visiting in the area of the Entertainment Centre at Glenelg. With the big crowd around, 8-year-old Clem becomes separated from his parents and siblings. His mother, brother, sisters and shop- keepers looked for him for hours. They searched in and around the shops and everywhere where they thought he might be. Eventually, the Police were called. The search continued into the night but being unsuccessful they returned home, intending to continue tomorrow. Then Esther went to get into bed but she couldn't because Clem was in it. After becoming lost he tried to find his parents but couldn't, so he headed for home and eventually found his way back past the zigzag post and crawled exhausted into the nearest bed. He must have past by those searching as he wasn't seen.

The railway line ran down the middle of the main street of Glenelg in those days and it was on one of these rails that Clem laid a penny to see what would happen to it when run over by the steam engine. In his excitement, he told his mother what he had done. She chastised him and told him that the engine may tip over and Clem was worried that he might be the cause of some disaster. Clem was so distressed that he may cause a problem with the train he did not even watch it, in case the derailment was blamed on him. However, there was no crash and a much relieved Clem recovered his coin to see it was only flattened a little.

Droughts and floods

At one time whilst on holidays at Glenelg, Ted got word of a very heavy thunderstorm that went through the district of Booleroo and further north. He and the boys returned immediately by train. Because of the rain, the area was mostly underwater with creeks and gullies running at full stream. The rail line from Jamestown to Booleroo was cut so they had to go by way of the Peterborough line then onto Orroroo. Some distance before Orroroo, the line was washed out over a gully and they couldn't go any further. They disembarked and walked into Orroroo and waited till the mail delivery wagon the following day to take them to Willowie. The watercourse between their farm and Willowie, Willowie Creek, which was usually dry, had burst its banks and the whole country was a sheet of water for as far as they could see. They were able to get home the next day and find that despite the heavy rain and some flooding, there was little damage. An interesting part of this return trip was that the day before they got news of the storm, Ted had purchased a pup. This little puppy was carried in the pocket of his coat for the entire trip.

Ted would keep hay and buck bush to feed his horses on during drought times but even then he had to destroy some horses due to lack of feed. Ted loved and was very proud of his teams of Clydesdale horses and loathed to have anyone else handle them. During extreme drought, he would put down the older horses rather than sell them to someone else and have them suffer further.

Droughts and dust storms were a real part of life in the northern agricultural areas in South Australia. On one occasion when the children were returning from school they could see this enormous cloud approaching at a fast rate of knots. At first, they were elated, thinking it was a rainstorm but this elation soon turned to despair as sand began stinging their legs and faces. They couldn't even see the horse that was pulling the buggy, let alone see the road but Daisy the horse took them home safely. When they got home and went inside their mother had candles and lamps burning to allow them to see. Sheets were used to cover the organ, piano and other furniture and sand-filled cloth sausages were laid at the bottom of doors to prevent dust from entering. By next morning the storm had past but imagine the mess there would be outside. They would be cleaning up for days.

Saturday night was always bath night no matter if you were dirty or not. A large oval-shaped galvanised tub with a handle at each end was used for the bath. This was placed on the floor and everyone took it in turns to have a bath, the cleanest person to bathe first. Some water would be added before the next person but this was one of the ways that water was conserved in those days. After everyone had finished with the water it was poured onto the garden and pot plants.

Some pet guinea pigs fell into one of the rainwater tanks at one time and drowned. The children, particularly the girls, were very sad at losing their pets. The guinea pigs were buried in the garden and wooden crosses were made to mark their graves. The water in the tank could still be used as the guinea pigs were in it for only a short time. With the water shortage, they could not afford to waste even this water.

As there was no radio or television to entertain them everyone had to make their own fun. Sometimes they would play cards. This was usually snap and similar games. Music played a large part in the family as Enid had studied well under Miss Jacobs of Booleroo Whim. She achieved her "Licentiate Music Australia" after the move to Western Australia. This allowed her to teach the piano which she has done for many years. Her mother used to play the organ for church on many occasions.

The boys also had plenty of entertainment by making their own fun. This was often in the form of practical jokes played on other family members and friends. Clem relates to one cold wet night, when Jim Wally, an employee, went out for the evening. Clem and Lance got an old oilskin rug, folded it tightly and put into the foot of the bed. When Jim came in and snuggled into bed, his feet touched this damp cold object in his bed, he yelled out, grabbed the sheets and flung them into the air and at the same time jumped clear of the bed. He thought he had encountered a snake in his bed. The boys thought this was great fun. It was the fastest they had seen poor old Jim move since he'd been there.

Prior to partaking of meals, grace was always said. But these mischievous children would tickle each other under the table whilst kneeling. The secret here was that you couldn't laugh or cackle as then their Dad or Mum would be down on them. A school friend was visiting the farm at one time and she commented on the closeness of the family and wished that her family would "read the Bible and say Prayers and be more like the Kentish's".

The boys got hold of some old pram wheels, about 400 mm in diameter, and fitted a stick into the hub for an axle. With one hand each side and their legs out behind them, they would run around and play with these for many hours.

Mrs Michaels, a neighbour, used to relate the children's activities to their parents and occasionally they would get into trouble for some of the mischievous activities in which they had been involved. So one day on the way to school they decided to play a trick on "The Old Tattle Tail " and they all laid down flat on the bottom of the buggy. As they went past where Mrs Michaels was, she could see the horse and buggy going along but could not see anyone in it. This would have caused her much concern and she couldn't help herself but tell their parents. Everyone had a good laugh as they had got their own back.

Another man that was employed on the farm was a fellow from Norway, Charlie Talkison. His forte was the curing and smoking of meat. Whenever a sheep, pig or beast was killed for meat some would be used fresh, and to prevent the rest from spoiling it would be cured in the smokehouse or corned to preserve it. Local wood was often used and with the spices and herbs added by Charlie, the taste would vary accordingly. After a period of smoking, the meat was stored in the cellar beneath the house for later use. Some of the meat was soaked in and infused with brine. This corned meat was also stored in the cellar.

Lance and Esther went to Willowie in the horse and buggy to do some Christmas shopping. They bought their purchases from Mr Wood's shop and as they were coming home they were looking at their parcels and some paper was rattling. This frightened the horse. As they didn't have hold of the reins, the horse lurched forward and broke the traces and then got out of the buggy and went off home. Here were two kids sitting in the buggy with dumbfounded looks on their faces not really knowing what had happened or what to do. They tried to push the buggy but of course, it was too heavy so then they pushed it off the road and began to walk home. As they began walking they saw their Dad coming across the paddock on horseback with the dog trotting along next to him. The pony had got home with some broken harness and he got on his own horse and came looking for them. They hitched up Ted's horse to the buggy as best they could and quietly drove on home.

Possibly it was in the early nineteen-twenties that Ted decided, much to the disappointment of the boys, to purchase a piano instead of a motorcar. Whilst in Adelaide he enlisted the help of Rachel Ellis to choose the LIPP piano. She was amazed as Ted just wrote out a cheque for the full amount. She would never have thought that Ted was that financial.

Sparrows or "Spoggies " were a real menace with crops, damaging more grain than they could eat. They would nest in the thatch of the roof on the sheds and stables. The boys were paid a bounty for each spoggy chick that they were able to kill. The boys would seek out and keep an eye on the nests, whilst standing on the horses back in the shed. You see there was a halfpenny for the eggs and a penny for the chicks (an ice-cream in a cone cost one penny), so being the enterprising young men that they were they could see a significant monetary advantage to let the eggs hatch before collecting them and claiming the reward.

Planting and harvesting

Wheat was the crop that was grown in this area. A team of 6 or 8 horses would be used to pull the plough and this was done in "Lands ". When using a disc plough this would form the paddock into low hills and shallow troughs. This was the accepted practice in those days. The shallow troughs would hold the moisture for the crop a little longer but this method did also cause some erosion. The hills would give the plants a good seedbed. After ploughing, seed and fertiliser would be broadcast by hand. A canvas bag with a broad leather strap would be worn around the neck or over the shoulder and about 40 Kg of seed would fill the bag. By taking a handful of seed and throwing it in a semicircular motion would spread the seed evenly. After the seed was spread, horses would be harnessed to a set of harrows and these dragged over the paddock to bury the seed to the required depth. Some farmers were using a horse-drawn seed drill but we believe that Ted still used the manual method.

Harvesting would be done (in the later years) with a horse-drawn mechanical harvester (probably a Horwood Bagshaw) with the grain being put into bags. The bags of grain would be loaded by hand onto a horse-drawn lorry (4, 6 or 8 horse team) and delivered to a rail siding or delivered to a shipping port. Lance told about a time when he accompanied his father on a trip to a port. Possibly Port Germein. Anyway, they travelled through the South Flinders Range on the main road. Lance was told by his father that it was important not to apply too much grease to the turntable of the wagon as on the steep and very winding downward grade the wagon might jack-knife. Should this happen, of course, the whole outfit would probably tip over. This did happen to some teamsters, according to reports.

Some crops were cut for hay. This would be done with a horse-drawn binder, which would cut the crop close to the ground and tie small bundles, each about 150 mm diameter and called " Sheaves " and deposit them in a line behind the binder. Men and women would come along after the hay was cut and pick up the sheaves and make a pyramid-like stack called a " Stook ". This would allow the hay to dry evenly. A team of horses would be harnessed to a lorry or wagon and people with pitchforks would load the sheaves. The wagon was driven to a suitable place and the sheaves were offloaded and formed into a stack. A good stacker could make a symmetrical stack and have the top arranged to shed any rainwater.

Money was extremely tight for most farming people and one way of overcoming this problem was for some shopkeepers and farmers to enter into an agreement of "bartering". Ted would take some eggs to a shopkeeper and exchange them for other items that he and his family needed for their daily needs. It was probably by bartering that they were able to purchase a bicycle, which the boys learnt to ride.

From the early 1920s to the 1930s there was the problem of "Diphtheria" among the younger children. The remedy to prevent this was to swallow a teaspoon of Kerosene. Of course, this tasted foul so the children would eat a quince to hide the taste of the Kerosene.

It was in 1924 that the first motorcar was purchased. This was a model "T" Ford and purchased for about, £ 200/0/0. This was complete with canvas roof and awnings, which could be put up when it rained. Windscreen wipers weren't fitted and sometimes a potato was rubbed over the windscreen to cause the rain to run off and improve the vision. In the winter chains were fitted to the tyres to get a better grip on the slippery road and tracks. Having a motor car made quite a difference to the family's activities particularly when going to church as they could get more done as the time taken in travel was less and also there wasn't the time taken to harness the horses or put them away and feeding after returning. Of course, some people had Chevrolets and Dodges etc. and these were more sophisticated but the model " T " Ford was a good vehicle.

About this time when Ted's younger brother Ernest was farming and growing potatoes near Mount Gambia. Ted had sufficient funds for the farm and his family's needs and honoured Ern's request for a loan to help him to plant his next crop. As Ted and his brothers were close, he helped Ern with his project by loaning him, £300/-/- that Ern used to plant his crop of potatoes. While there are no records of this loan being repaid, there are some conflicting memories of repayment. Ted wouldn't ask for it but the money would have been very handy with their move to Western Australia.

Changes for the better

It was probably because of the droughts and the fact that he didn't want his family to suffer the hardships that he had, that Ted decided to check out the promise of land in Western Australia that might better suit their purpose. Ted had had several good seasons and he expected them to come to an end and thought that this would be an opportune time to move. Many wiseacres said that taking his wife to a wet climate like the South West of Western Australia would kill her. She suffered badly with asthma, particularly during the dusty summer. However, her health was much better for the change. With Ted being 61 years of age had a tough time making the decision to make this move and spent many sleepless nights weighing up the pros and cons.

Before looking for land, Ted arranged with a neighbour, Mr Suffield, to purchase the 640-acre Oldham Farm from him at a price, £5/-/- per acre should he be successful in finding a suitable property. Knowing this, Ted was able to budget on the amount he could spend to purchase another property.

In early February of 1926, Ted travelled by train to Perth at a cost of, £7/-/- return, where he met with a Mr Halbert Service, a land agent, and began the search for a suitable property. He got word of a property South of Perth that was for sale and went to investigate. The Hall family owned this property.

When Ted arrived at Bolinda Vale for the first time he was completely taken with the available water. Being in the height of summer he was enthralled with the green grass near the creek, the green scrub along the creek banks and of course with the water flowing in the creek. This was a complete contrast with what he was experiencing at Booleroo Whim.

Ted had a discussion with a Mr Matthews who informed him that he had done well and brought up his eight children on his farm in the district. This discussion and others with the Halls and other neighbours in the district caused Ted to form the opinion that Bolinda Vale near Keysbrook was the place for him and his family to start with. Once here he could look around to find something better. What Mr Matthews didn't tell Ted was that he was also drawing a wage from the Government as a Group Settlement Supervisor in addition to farm income to support his family. Ted didn't find this out until some months later.

Purchase documents were drawn up and a deposit paid. Ted returned by train to Port Augusta or Hammond and then on to Booleroo Whim to finalise the sale of Oldham Farm, collect his family and belongings and relocate to Bolinda Vale farm at Keysbrook in Western Australia.

The property deal was completed with Mr J W Duffield in March of 1926. After the farm was sold, a clearing sale was held to dispose of all of the equipment that wasn't economical to take to Western Australia. Some household furniture, beds, mattresses etc., all of the livestock and the woodpile were sold at this sale.

The family moved to stay with their friends, the Thisseltons who had recently moved to Yandiah, with Esther at the Nicholls' Hope farm and Clem at Greig's (a neighbour) for a week until they were ready to board the train.

Initially, Edie was not entirely happy with the move to Western Australia, as they would be leaving behind all of their family and friends. Some people had told her that the wetter climate over there would not be good for her asthma. This would have added to her anxiety about the move.

Enid cried when the Farm was sold. She is the oldest child had settled into a way of life and was loathed to see this changing. But despite this, she had a secret excitement about the move and all the adventure that goes with it. She was a little disappointed because she would need to wait a further year before she could obtain her motor vehicle driver's license.

Lance was a good scholar, had passed his school exams and achieved the Q. C. (Qualifying Certificate). This qualified him to receive a scholarship or bursary to attend the University in Adelaide to further his education as he had aspirations of becoming a doctor. But with the family moving to Western Australia, he had to forgo this chance to continue to improve his education. He was disappointed with the fact that he wasn't able to continue with his intended education but this was probably offset by the excitement that he could envisage in the near future with the move and the making of a new home and friends.

Esther was very excited about the move and being only just nine years old, this move was the event of a lifetime. The leaving the old place, the boarding of the train, waving good-bye to friends and relatives, the two or three-day train trip, a few days in Perth and then finding their new home. These things were going through her mind and her excitement was at a peak with this new adventure.

Clem was also excited at the adventure of the shift and having a new property, but he was also very disappointed that he was leaving behind his very good friend, Cora. She was the daughter of the local blacksmith, Mr Thisselton, who had recently moved to Yandiah. She and Clem had been very good friends and attended school together. Clem and Cora continued their friendship by correspondence for many years after the move to Western Australia. Their mothers also kept in touch as they had formed a solid friendship during their time at Booleroo Whim.

Aunty Alice, Ted's sister, was very disappointed that they were selling the old family home and property and moving all the way over to the West. She tried to talk Ted into taking up Uncle Ern's suggestion of moving to Mount Gambia to be closer to the other family members and a better district for farming. She wasn't at all impressed with Bolinda Vale at first but after a few years, she mellowed and also became to respect Ted's decision to make the move.

The Kentish family had been at Oldham Farm near Booleroo Whim for about 50 years, and being involved in Church and district affairs for most of this time they were sorely missed by the other members of the community. Before they left to stay with Thisseltons, the community gathered together to offer a going away gift to Ted and his family. A silver tea set was presented to them and Ted received a Memorial Scroll (a computer copy of this is included below).

The family had a very good and faithful dog named "Dash" and he was given to one of the neighbours to keep. When Ted returned to collect the last of their belongings, the old dog had returned to Oldham Farm. Ted decided that seeing as how the dog wanted to stay with the family, he paid to have the dog travel with them on the train to Perth. The old dog went bush soon after arriving at Bolinda Vale and wasn't seen again, much to the disappointment of the family.

The Nicholls Story

Joseph Bell arrived in Hobart Town in Van Dieman's Land (Tasmania) on April the 7th 1833 where he was employed as a builder. On December the 23rd 1834 he purchased a property and on February the 28th 1835 he married Ann Patton.

After living in the Town for 2 years, the convicts, bushrangers & "ticket of leave" men were causing many hassles for them & they decided to check out this Adelaide in South Australia. Joseph arrived in Adelaide aboard the "EUDORA" in November the 27th 1837. After erecting his weatherboard house, he returned to Hobart and brought back his wife and 2 sons aboard the "ABEONA", arriving on March the 23 rd 1838.

The earliest record of Joseph having land was dated as of January 1842 with the sale of a 1-acre lot at No 92 Fuller Street at Walkerville. About 1 Km from James Kentish's (Snr) farm at Walkerville.

Joseph continued with his building, also taking on the task of Undertaker for a short period, until about 1840. In 1841 Joseph acquired an 80-acre section No 175 of STURT near Glenelg. About 5 Km West of the Kentish's RYE FARM at Goodwood. This he named "COBHAM", where he began his farming activities. This was virgin bushland and all clearing was done by hand with the help of horses. By 1842 Joseph had 30 acres of land planted to wheat, 6 acres in barley and 2 acres in oats. In addition to this, he also had vines, fruit trees, figs and almonds. During the mid-1840s a Reaping machine was used by Joseph to harvest his crops. This was horse-drawn and could harvest about 7 acres per day.

In January of 1859, Joseph purchased 500 acres, being No. 10 in the Kapunda District, County of Light for, £1290/-/-. This was farmed by his son Joseph (Jnr) until his marriage to Elizabeth Ann Puxton in 1865 when he then leased a property for himself near Waterloo in the Mount Lofty Ranges, about 110 Km East North East of Adelaide.

Mary Ann Bell was the first Bell child born in Adelaide. She was born on 27 of February 1839 in the first stone house to be built in the city, situated in West Terrace. This was just before the family moved to their new home at COBHAM Farm.

There are not any records of Mary's junior years but I imagine she would have been busy assisting her mother with keeping house and helping with the younger children, as she was 13 years old when her mother's last child was born. At the age of $19^{1/2}$ she married James Nicholls in her family's home at Cobham Farm. This would have been a fairly colourful affair with her younger sisters attending her at the ceremony. The Reverend William Nicholls (believed to be the Groom's uncle), an independent minister officiated at the ceremony.

The congregation at St. Peter's Church, Glenelg to which the bride belonged, presented her with a family bible, which is now in the possession of James Nicholls 3rd.

They made their home Brighton (just South of Glenelg), Kapunda (about 80 Km North East of Adelaide) and Golden Grove (about 20 Km North East of Adelaide). Mary Ann was the church organist for the years that she resided in these areas as well as bearing and rearing their first 7 children.

In 1870 they moved on to Red Hill and from here to Booleroo Whim in 1876 where they took up Section 23 of the Hundred of

Willowie, which they named "HOPE FARM". This was situated about 5 Km West of Booleroo Whim and about 10 Km by road from Oldham Farm, which would be taken up by David Joel Kentish & family some three years later.

Upon arrival at Hope Farm, the children would have been about the following ages: James (Jnr) 17 years, Annie Louisa 15, Julia 12, William 11, Elizabeth 9, Alfred 8, Kate Emily 6, Marion 5, Ada Amy 1. Edith Alice, Harry & Grace were born after the family arrived at Hope Farm.

Until the opening of the school at Booleroo Whim in 1880, the Nicholls children were lucky enough to be taken to school at Merle's in a covered dray. Walking to school in those days was the norm. One of the girls had a job in Booleroo Centre and she used to walk there on Sunday afternoon and board during the week. On Saturday she would walk back to the farm for the weekend.

Two of the boys attended The Stanley Grammar School in Watervale for at least one year each. William attended in 1881 at the age of 16. James attended in 1885 at the age of 25. The cost of this education was 13 Guineas per quarter paid in advance for full boarders.

There is a story of the large oven that was constructed on Hope Farm. With rocks and mud, the sides and back were built up to about chest height, then curved sticks were laid on top of these walls. Mud and stone were plastered onto these curved sticks and after a period of drying the fire was lit. The fire would cook and set the mud, burn out the sticks and if the correct mud was used they had an enormous oven. This was used to cook mutton (probably whole for a family of that size), bread, pies and probably yeast buns.

1876, the year of their arrival was a drought and most farmers were carting water for stock and domestic use from the Government well at the crossroads. A whim was fitted to this well, which lifted the water from approximately 30 metres to the surface. Buckets were fitted to an arrangement of cables, which were in turn attached to a winch rotating around a vertical pole. Cross arms are fitted to this vertical pole and a horse was attached to operate the Whim. He would walk forward and this would raise one bucket while the other was going down. When the full bucket reached the surface, it was emptied. The horse was turned and walked in the opposite direction to lower the empty bucket and raise the other, now full bucket. At about 4 hour intervals this horse was rested and another harnessed and used.

Kids, even in those days were inquisitive. As Harry was watching the Whim in operation and leaning over the side of the well he slipped and fell. There is no mention of injuries but to fall 30 metres, that water would have been awful hard. This happened in 1880 and he went on to live until he was 62 years of age.

The Bell and Nicholls families have always had a close relationship with their God and church. Mary and James were no exceptions to this. Both of them were good singers and welcomed to the Church Choirs. Church services were held in the Booleroo Woolshed until the congregation banded together and constructed their own church on land set aside for that purpose. This church was to become the focal point of the entire community. Many a night would see meetings, lectures and debates taking place in the church and on Sunday after morning service probably a communal picnic.

It was the attendance of Edith Alice to the Sunday School at the church which led to her meeting Edward Joel Kentish. They were both good Christians and avid churchgoers with good voices.

Ted saw her lifting a bag of sugar onto a horse-drawn buggy and told his mate "She looks like a good wife for me"

Edith Alice Nicholls was married to Edward Joel Kentish on February the 20th 1908.

South Australian property details

"Bent St. property " A one-acre section within the Adelaide Town Mile being number 98 Grenfell Street, was purchased by James (Snr) in 1838 for, £65/-/-. This was situated at the N W corner of what is now Grenfell and Frome Streets. Bent Street divided this lot in the middle.

"GOODWOOD FARM" between Goodwood & Unley Rd Goodwood, about 5 Km South of Victoria Square in Adelaide, in the name of James Kentish (Snr). 39 acres of this was gifted to James (Jnr) on June 2nd 1846 in lieu of, £110/-/ for work done on the other farm. This was the beginning of "RYE FARM". A 41-acre part of it was transferred to David Joel in lieu of payment of, £120/-/- on 29th February 1846. This was sold on the 31st March 1857 to his brother James for, £700/-/-.

"WALKERVILLE FARM" Section 478 on the River TORRENS, east of Lansdowne Terrace, approximately 5 Km North East from Adelaide centre. This was a James Kentish (Snr) family farm.

David Joel took up section 4246 of the Hundred of Munno Para using a land grant of, £84/-/-. This was 80 acres and was sold 20 months later for £120/-/-.

 A land grant allowed David Joel to take a section in the Hundred of Saddleworth in 1855. This was sold in 1857, for £595/-/-.

"PROSPECT FARM" comprising of 4 sections, 3313, 3314, 3315 & 3330 of the Hundred of Munno Para totalling 348 acres, in the hills of Gawler approx. 40 Km North- North East of Adelaide.

This was David Joel and family's second home. The purchase price was £1000/-/-. Sold in 1879 for, £1540/-/-.

541 acres in Sections 3 & 6 in the Hundred of Appila. Which is approximately 210 Km North of Adelaide. This was leased on 21 March 1894, by Ted and his older brother Herb, from their Uncle James for a period of 7 years. This property was just north of Uncle James Kentish's farm "THE PINES".

"Yandiah ", about 15 Km South of Booleroo Centre. Probably between 1901 and 1903 Ted and Herb went mining for Silver and raised, £100/-/- to help with the purchase of land.

"OLDHAM FARM" sections 73 & 79 of the Hundred of Willowie, approximately 250 Km north of Adelaide. David Joel purchased this farm in 1879. Probably named after the town his wife lived in as a child & young lady in Lancashire, England. Willed to Ted upon the death of his father in 1903. He was to pay the other surviving children an equal share and a trust for the children of his eldest brother Henry. This farm was sold in 1926 when the family moved to Western Australia.

Maps in South Australia

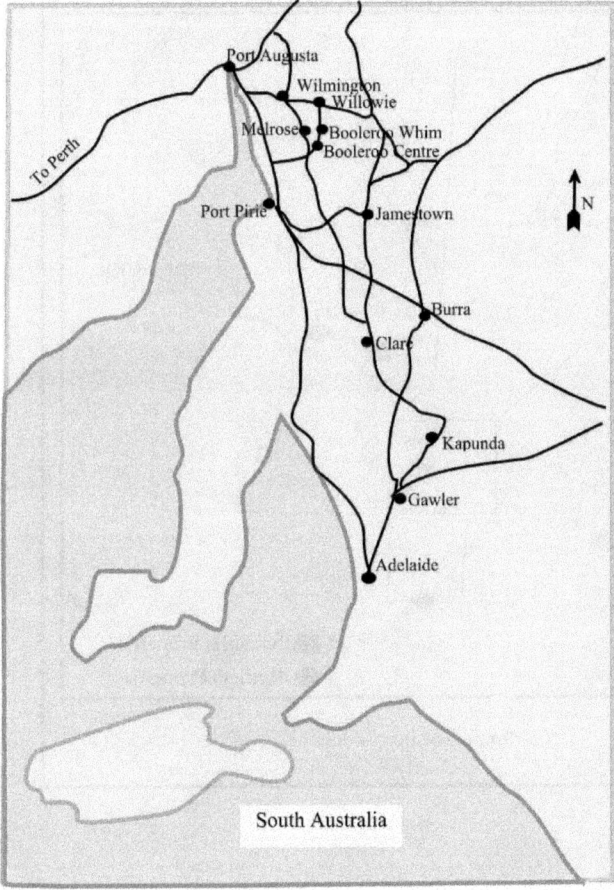

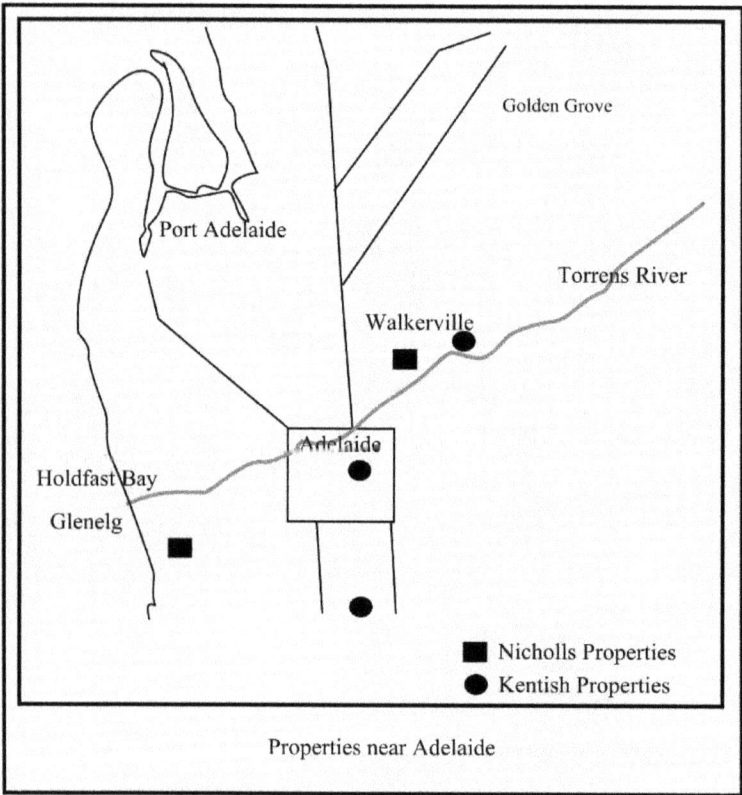

Golden Grove

Port Adelaide

Torrens River

Walkerville

Holdfast Bay

Glenelg

Adelaide

■ Nicholls Properties
● Kentish Properties

Properties near Adelaide

A Memorial to

Mr. E.J. Kentish

From the residents of Booleroo Whim
and Willowie Districts

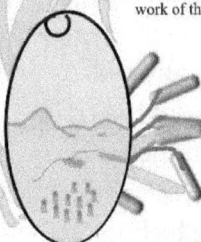

We, the undersigned, bear testimony to the splendid services rendered by Mr. E.J. Kentish to the above districts. He splendidly and loyally served God in membership with the Methodist Church at Booleroo Whim, holding all of the official positions a layman can fill, i.e.; Circuit Steward, Lay Preacher, Trustee, Sunday School Superintendent, Vice President of the Wesley Guild and Band of Hope. He preached effectively throughout the Melrose and Booleroo Circuit and at Willowie.

He was a ready supporter of Missions and all other worthy objects, contributing freely and also responding readily to every public demand made upon him.

He exhibited much interest in the Literary Society at Booleroo Whim and took an active part of the Willowie branch of the Agricultural Bureau.

His influence was widely felt and he was a great power for his righteousness amougst all who knew him and his character was irreproachable, his ideals high and he rendered the community every service in his power, his best service being in the work of the Church.

Signed:~
GW Shapely for Wilmington Circuit.
Bertram S Howard for Melrose and Booleroo Circuits.
WH Hughes for Old residents.   F Duffield for Old Residents.
N Sime for Willowie Church.   H Slewdys for Sunday School.
Wallace P Foulis for the district. Geo Michael for Church activities
SG McCallum Kindred Societies.   AJ Schmidt for Whim Trust
John F Linklater for Kindred Societies.

March 1926

Western Australia

The family, Ted, Edie and their children Enid, Lance, Clem and Esther, travelled to Port Augusta with their hand luggage in their Ford "T" car. The organ, piano and crates of clothing, furniture and other items went on ahead of them by a carrier. At Port Augusta, their belongings were loaded onto the train. The crates and furniture in a covered rail wagon, the dog in a crate in the Guards' Van, the family in a sleeper car and the car on a flat top wagon. Because they lived frugally, they would have shared their accommodation whilst on board the train.

To give some perspective to their ages, in the year of 1926 they would turn the following age: ~ Ted 61, Edith 49, Enid 16, Lance 14, Clem 12 and Esther 9 years.

As the train left the station they waved goodbye to their relatives and friends that had come to see them off. Although there were tears they were all excited at the prospect of the new adventure, which lay ahead. As the steam engine gathered speed so did the boys and Esther as they searched out the train to see what was where. Although it was late March they didn't notice the heat as they headed across the Nullabor, this was probably due to the exciting elation that they were feeling. Ted and Edie would have had some quiet time together to discuss plans for their arrival. Enid didn't enjoy the trip across as much as the others as she was affected with motion sickness and wasn't too well. She spent much of the time in her bunk.

The train made regular stops while crossing the Nullarbor and the passengers had several encounters with groups of Aborigines at most of these stops. A white woman by the name of Daisy Bates was in the desert working with a group of Aborigines and she was with them at this time. They were asking for donations of "white mans' money". This meant that they wanted money to help with their education and living but only if it was in the form of silver coins. They didn't want copper coins. Most of the passengers did make donations and these would help Ms Daisy Bates with her work with the Aborigines.

When the train arrived at the station of Kalgoorlie, everyone and everything had to be offloaded and reloaded onto another train. This was because of the different gauge of the railways. From Port Augusta to Kalgoorlie they travelled on the "Standard Gauge" line and from Kalgoorlie to Perth they were on the "Narrow Gauge".

At this stop, the two boys and their father took the "T" Model Ford car and drove the rest of the way to Perth on the rough dirt track that was to become "The Great Eastern Highway". They made fairly good time to get as far as Coolgardie and be on the station platform to wave to the rest of the family on the train as they went through the station. Because of this rough track, Ted and the boys took several days longer to get to Perth than the train.

As the train was entering the area near Northam, Edie and the girls were surprised at the vast change in the countryside. From the rolling plains of drying bush with occasional farming land giving way to the undulating bushland of the Darling Escarpment. This would be a very welcoming sight and showing a vast contrast to the land, which they had left behind in South Australia. Even though the bush was fairly dry at this time of year they could see some of the wildflowers that they had heard about and marvel at their beauty and diversity.

When the train pulled into Perth station they were met by Uncle Alf, Edie's elder brother. Uncle Alf had his house on the corner of Hill Street and Adelaide Terrace in Perth. He was farming the area and had his dairy cow grazing on the Esplanade, which is the land between Adelaide Terrace and the Swan River. Edie and the girls stayed with Uncle Alf until Ted and the boys caught up with them. They spent these few days looking around Perth and the Swan River; saw their first black swans and the other wildlife that abounded around the river's edge.

Ted and the boys finally arrived in the Ford "T". The road trip had taken a little longer than expected but they weren't concerned with this as they had seen much more of the bush and the areas east of Perth. The rough dirt track continued from Kalgoorlie to as far as Northam where it, although not great, was a better road and macadamised too.

After spending a night or two with Uncle AFL they loaded themselves into the car and headed south towards their new home. The road as far as Cannington was in fair condition as this was the southern suburbs and the beginning of farming land. From Cannington to Keysbrook and further, the road was not formed but only a corrugated, rough minor dirt road. During the trip from Perth to the farm, Edie and the children were getting increasingly impatient and asking, "Is this it?" Ted had told them of the hill, which was the landmark of Bolinda Vale.

As they approached the property, with about one and a half or two miles to go, Ted pointed out the hill, which they could see in front of them and said, "That's our hill. Right there in front of us". Everyone was pleased that the long trip was nearly over but their continued excitement kept them on their toes. Just before they crossed Dirk Brook, the creek that runs through the property, they took their first look at what was to be their new home and farm, Bolinda Vale. There was water flowing in the creek. There was a fair amount of green grass in the paddocks and also amongst the scrub on the creek's banks, where it meandered around close to the road. A few hundred metres after crossing the creek they turned from the road and drove up the track to a group of buildings. One of these was the homestead, which still stands today, with the verandah all round and a large grapevine on a trellis at the back verandah. Apart from the house with an outside washhouse and an old creeper covered workshop close by, there were stables with yards for the horses, a cow yard and a small tool shed. Most of the outbuildings were fairly ramshackle affairs.

This old workshop was, in fact, the original living quarters of the Hall family prior to the building of the Homestead. Mrs Hall was installed here as a bride. Some of the iron of the roof of the Homestead still has the shipping markings of "E A HALL Old Keysbrook ". Old Keysbrook, of course, referring to the old Keysbrook Railway Station, which was across the highway from Bolinda Vale, close to where Coralie and Graham now live. This being 37 miles and 38 chains (60 Km) from the main Station of Perth. It wasn't until 17th November in 1913 that it was moved from this position to the present site of 39 miles and 6 chains (62.8 Km) from Perth Station. The Balgobin Station (near Page Road) was closed at this time also.

Ted and family kept in contact with Mrs Hall after she left the farm and resided in Guildford.

Practically before the car had even come to a stop they got out from the car and carried out their own inspection of their new home and farm. The items of greatest interest were the creek with the flowing water, the massive bush-covered hill out the back of the house to the east, the small orchard with navel oranges and an apricot tree that was behind the house and the big windmill that was pumping water from the creek to a large tank on a high stand, just out in the paddock from the house. These would contrast greatly from the surroundings of that which they had left behind them at Oldham Farm at Booleroo Whim.

The lavatory was an upright structure made from bush poles and corrugated iron and a swinging door probably with a crescent or star cut into the door. This was situated some 30 metres out the back of the house just past a large Red gum tree. With the lavatory being so far down the back you needed to plan your trip so as not to be late for your appointment. It was a fairly modern affair with a pan and classed as a "one holer thunderbox ". In later years this was demolished after another weatherboard "one holer thunderbox" with a pan was built just out from the washhouse. This was a big benefit as being much closer to the house, less planning was necessary when one needed to visit the "dunny".

A few days later the other items that travelled with them on the train arrived at Keysbrook Railway Station. They collected these and returned to the farm. On unpacking, they could find only one item that was damaged. It shows that much more care was taken in those days in the handling of goods during shipping. One of the old Oregon timber crates that were used is still in the old house. Someone had modelled it into a dressing table.

Accidents didn't play a large part of the family's life but soon after arriving, Edie was putting up curtains on one of the windows when the wooden box that she was using to stand on, collapsed. She fell to the floor and broke her arm. Ted took her to the nearest Doctor, who was at Pinjarra, in the model "T" Ford along the rough road. That trip would have been fairly painful as the track was fairly rough and the suspension of the car was a bit harsh. However, the Doctor applied plaster to the arm and advised them to return home and for Edie to take it easy for several weeks. You can imagine her doing that! She was soon back doing the housework and Esther remembers seeing her doing the washing with one arm out of action. She would be using a copper to boil the linen and a washboard for scrubbing the dirty work clothes. Quite a job with two good hands, let alone with one out of action.

Although Ted had been a wheat and sheep farmer all his life, he made the commitment early to get into dairying. They purchased some dairy cows, mostly Jerseys, and this was the main source of income for the farm. At the start of the dairying, the cows were milked in the yard, as there was not a dairy building as such. Ted, Lance, Clem and Esther would each milk 5-6 cows by hand in an hour whilst sitting on three-legged stools. The milk was poured into ten-gallon (approximately 45 litres) cans and then cooled.

Esther would make a game of all the jobs that she did. This not only relieved any boredom but also made the job fun. She would also use this method to increase her pace and therefore get more done in a day.

The cooling of the milk was difficult as there was no refrigeration. At first, the cans were attached to a pole across the creek with a rope so that the cans were in the water. This was fine until it rained heavily and the cans got washed away in the flow of water. Later a cooler, known as a Coolgardie Safe, was built near the back verandah of the house. Under the grapevine trellis, they constructed a framework of timber and thatched the top and sides of it. Water was trickled over this thatch and as the breeze flowed through the thatch it cooled the air. Not only was the cream kept here but also the butter and other food items that needed to be kept cool.

A hand operated cream separator was used to separate cream from the milk. A vat on the top of the separator was kept filled by one person, while another turned the handle. So as to get the cream to an even consistency, it was very important to turn the handle at exactly the correct speed. Lance was usually in charge of the separating and he had the speed correct and would make sure that the person turning the handle was aware of the speed. Sometimes a bit of verbal encouragement was required to accomplish this. For the most part, this cream was taken to the Keysbrook Station by a Mr Ted Furber on his truck and then loaded onto the train to take it to Perth for sale. He would usually call in three times each week and take two to three cans of cream each time. Eventually, Mr Les Bee gained the cartage contract, using one of Pascomi's trucks and he would cart the milk and cream to the City each day, so relieving the cooling problem.

Some cream was kept for their own use. The cream was never short on the dining room table. Butter was also a by-product of the milk. This would be churned and pattered by hand and it was the best tasting butter for miles. Edie's yeast buns and their butter were a great combination and drew many a visitor back.

The skim milk, that's the milk, which is recovered from the separator during the process, was fed to the pigs. These were running in the paddock to the east of the house, sometimes across the creek. These would be used for their own consumption with some being traded for items under a bartering system, and some were sold.

For the first year after their arrival at Bolinda Vale, Clem and Esther were educated at home by correspondence due to the recent closure of the local school. Mrs Middleton had told Edie about education by correspondence, much to the dislike of the children. The boys' old room was converted into a classroom and it was here that Clem and Esther continued with their education, with help from Enid.

The old school building that was at Balgobin was shifted to Keysbrook. (The Balgobin School was situated just east of Bunbury Highway and north of Page Road. A group of pine trees marks the area). Clem, Esther and about ten other pupils attended this school. In the 1930s this school closed again due to low numbers of students. At first, they travelled by horse (Bobby) and sulky (two wheels), taking the Wallace's (neighbours) children with them and later when the Wallace' children were old enough to have their own ponies, Clem rode his pushbike and Esther rode her horse. There was a hitching rail near the schoolroom where the horses were tethered while the children were in class. Esther or Clem would water the horses at lunchtimes. During the winter the horse would get quite cold even though it was covered with a rug and when it was harnessed to the sulky it sometimes would rear up to show its displeasure. A few pieces of harness were damaged when this happened so they did have to be very careful; otherwise, it was a walk home.

When Esther was about 14, the Keysbrook School closed and the students were required to attend the State School at Serpentine Bridge. Esther preferred to attend college in Perth and do a course in bookkeeping but Ted decided that she was too valuable on the farm and would learn more anyway, so he requested her to forgo her education. Although she was disappointed with not being able to further her education,

Esther enjoyed the farm and the farm work and was quite happy to stay.

At about this time Esther and Enid were given a pushbike to share. This was a very useful method of transport and offered a change from horse riding. Most weekdays one of the girls would ride into Keysbrook to collect the mail from the Post Office. Ted would sometimes collect the mail with his ex trotter called "Bouncing Dorothy" in the shafts of the sulky. It was quite a sight to see this horse trotting along with its' high step. Esther would sometimes go to collect the mail with the horse and sulky and frequently travel with a lamb on the seat beside her, much to the amusement of the locals, who would comment " Here comes Mary with her little lamb ".

Some of the riding horses were Brumbies that had been trapped in the hills, These made excellent riding horses, as they were very robust and stocky and suited to the hard terrain of the hills. One of these was named Pharlap and another Trixi.

When they purchased the property, they were fortunate in having inherited a young Aboriginal by the name of Mickey Ball. Mickey was a full-blooded Aboriginal and one of a twin. His mother was from the Nyngan tribe who feared the superstition of giving birth to twins. To alleviate this superstition she abandoned one baby and disappeared with the other. Mr and Mrs Ball took Mickey in and he stayed with them on their property at Laverton and later at Bolinda Vale. Ted would teach Mickey to read by giving him a bible and working with him through the 23 rd Psalm. He was quite a hard worker around the farm and was a great sportsman. Mickey, Lance and Clem were great mates but sometimes he would take off on walkabout with his dog.

Occasionally one of the sulky wheels would rattle. These were constructed from wood and were fitted with a steel tyre. As the wood dried out the tyres would become loose and sometimes rattle. Esther remembers Mickey saying, " Don't worry Mr Kentish, the tyres won't come off, they will only rattle ". Sometimes these were put into water to swell the wood to keep the tyres tight.

Mickey returned home ill from one of his walkabouts and developed pneumonia. He was nursed at the farm for some time and they advertised for a nurse to take care of him. A nurse was found and arrived at Serpentine Station expecting to find Mickey Ball in an Aboriginal humpy but was surprised to find him bedded down at Turner Cottage near the Serpentine River Bridge. Mr Hardey collected her from the station and took her to Turner Cottage. Here she nursed him for several days until his death in 1935. As there was no preacher in the area at the time, Lance conducted the burial service at the Serpentine Cemetery. To mark the grave they brought down from the hill a granite boulder and had the retired stonemason, Mr Oliver Drake, carve in it "MICHAEL BALL an AUSTRALIAN GENTLEMAN ". This headstone is still in place today.

The reason that Mickey Ball was nursed in Turner Cottage was that the people who adopted him, the Balls from Laverton who later owned Bolinda Vale, had a daughter (she and Mickey grew up together) who married into the Turner family and they made this cottage available to Mickey for this purpose.

Enid received the best education of the four children and she would travel to Perth by train once a month to study music and attend music tuition by Mr Lecky at the cost of, £0/12/6. She continued with her music teaching on the farm. Enid would do some "fill in" work on the farm and occasionally help her mother with the housework but spent many hours each day practising her piano playing and study for her degree of music (Licentiate of Music, Australia) which she achieved on October 16th 1936. After the evening meal, Enid would say to the others "You do the dishes and I will sit at the piano and play to Daddy". She had the use of an Austin Seven motorcar that she used to travel around the district giving piano lessons. Later when her father passed away this became hers. Although Enid did not have a lot of physical input to the family farm, she did provide her parents with spiritual benefits with her music.

Edie was a great cook and always busy at it. The table was always covered with an abundance of good wholesome food. She would bake her own bread; make her own butter, jam, and other preserves. There were quite often visitors for meals. I think that with the food and companionship being so good that many people adjusted their timetable so that they could be there at mealtimes. The usual jocular comment regarding the quantity of food at the table when excess visitors arrived was "We can always add more water to the soup ".

The family continued with their Christianity, holding Church services in their home on most Sundays. Many of the local people and neighbours would attend. The service would sometimes be conducted by visiting clergy but Ted was a proficient lay preacher and would do a good job when they were not available. Hughie Manning recalls many happy hours spent at these services and fellowship. They were usually followed by Sunday dinner. Clem and Hughie sat close together at the table and shared the food of the extra plate that Clem always seemed to procure.

A Minister was stationed at Jarrahdale for some years but he left and then the district relied upon its own resources for spiritual contentment. Clem would often ride his pushbike from Bolinda Vale to Jarrahdale to conduct the service in the church. On his way past Manning's, he would collect Hughie and they rode together.

This family home fellowship was to continue for many years and was a catalyst to several lasting friendships and marriages. These services were later to move and be held in the Keysbrook Hall. This created some disappointment for Ted, as his home was always open to those who wished to worship with them.

Prior to each meal, grace would be said with everyone at the table kneeling on the floor. When the morning meal had been completed Ted would kneel and pray, then read from the Bible. Edie would then sit at the table and read from her prayer book while everyone knelt. This is the way they conducted themselves and each of their children continued on with this practice in their own homes.

The creek, Dirk Brook, flowed fairly fast until it reached the level flat area near the house. Here it meandered around a great deal as it found its own way through to the Bunbury Road. There were two main places where this meandering took place and during the winter the swollen creek would burst it low banks in these areas and cause some local flooding.

Edie used to have good flocks of geese and Muscovy ducks. These would provide feathers for pillows and mattresses as well as meat for the table and sometimes income from their sale. Sunday dinner was usually roast Duck served with roast potatoes, beans and pumpkin, finished off with bread and butter pudding with cream for sweets. For the most part, the birds were free-ranging but with shelters into which they were occasionally locked to save them from the foxes. The Muscovy ducks would nest along the banks of the creek amongst the scrub and tea tree. During the first few winters, most of the nests and eggs were washed away during storms in the water flow but only a few ducks were lost.

Ted decided it was best to stop this flooding by straightening the creek. To do this they chose areas of the creek where there were the shortest distances between the meanders. They would use picks and shovels to dig a narrow trench to join up the two areas and by banking the water to the meander would cause the water to flow through. As the water flowed they would use crowbars to break away the sides of the trench to increase its width. This method was used in several places to improve the flow of water through the creek to allow the water to flow away and so to prevent the flooding. While this was quite acceptable at the time and did reduce the flooding, it has been proven over time to have been the wrong move. While the creek was flooding and meandering, it was filling the ground with water, which was very beneficial for the summertime. Now that the water flows away more quickly, the groundwater has dropped and the area around the creek is much dryer. It would have been more beneficial to put up with the short-term problem of floods than to have this lack of soil moisture now. Unfortunately, we don't always see the full effect of our actions until much later, when in many cases it's too late to rectify.

Fruit was not grown in Booleroo and the orchard was one of the saving graces of Bolinda Vale. Edie would take forward orders for their fruit and this would be dispatched just after picking, usually by train. Some were sent as far as Kalgoorlie. The orchard that existed when they took on the property was only fairly small, so Ted decided that they should plant more fruit trees and bring the orchard up to commercial size. He selected the area where the creek used to meander. Now that it didn't flood too much in the winter, this was a fine site for the orchard. Whenever a horse or a cow died or was put down, he and the boys would bury it in this area. The rich alluvial soil was ideal for the fruit trees so they planted Navel Oranges, Valencia's, Grapefruit, Mandarin, Apricot, Loquats and more, over an area of about 6 acres. For most of the year, there was plenty of moisture in the ground for the fruit trees but during a long hot summer, they did need to irrigate from the creek to keep the trees in top condition.

With the variety of types of fruit trees in the orchard, the fruit ripened at different times and this enabled them to pick the different fruit about 2 or 3 times each year. Picking was done by hand with the picker wearing the strap of a large canvas bag around their neck. This bag had a flap attached to the bottom that allowed the fruit to empty out gently so as to not bruise it. Once the fruit on the lower branches had been removed they used to stand on the side of the dray, which was harnessed, to a horse. Provided that the horse stood still, the picker did his or her job very well but occasionally the horse was restless and some time was wasted in repositioning the horse after picking oneself up off the ground. Ladders were also used but these didn't give the same sense of achievement as the horse and dray. Esther had a pony with a large bag attached to each side and this was very useful when picking fruit. Once the fruit was picked it was sorted into varying grades of size and packed into boxes. During the war when labour was hard to get, a fruit grader was bought and used. This reduced the time taken for grading and packing considerably.

A dump case would hold 1 bushel and a flat case would hold 3/4 of a bushel. These cases of fruit were forwarded to the markets in Perth for sale and for delivery to the customers who had ordered them from Edie. Each morning the windfall fruit would be collected and fed to the pigs. This practice kept the pigs fed and most of the fruit flies away and by spraying with a mixture of tobacco leaf soaked in water frequently, would keep them and other pests under control. Tobacco was fairly cheap and easy to obtain until the Government began controlling its production. Then other, more expensive products were needed to control the pests.

During the depression years, there were many tramps, swagmen and bagmen travelling through the district. It was a very common sight to see them doss down in the hay shed for the night. Most would get a meal for doing some light work such as chopping some wood for the fire and all were made welcome to stay for the night in the hay shed, provided that they observed the rule of NO SMOKING.

An unusual event occurred when at the evening meal table with one of the tramps that were befriended by Ted. The tramp mentioned that he knew of a stolen car that was down by the railway line, so down they went to have a look at it. Although the news of stolen cars is quite common these days, it was most unusual then. They found the car and reported it to the Police and discovered later that the tramp, who told them of it, was, in fact, the person who had stolen the car but he was, by this time, long gone.

Working and clearing the farm and operating the dairy took a large part of each day but they seemed to find time to go on picnics and many other activities. One of these regular picnics was to pack a lunch and climb the hill to the east of the house and walk through the bush to an old disused timber mill where an old house stood by a small flowing creek. It was a delightful walk through the bush, particularly in the spring when all the wildflowers were out. The wildlife was abundant with possums, kangaroos, wallabies, brushes, birds of all sorts including scrub turkey and wild duck, the occasional red deer and many of the smaller animals which run around on the ground of the bush. After the walk through the bush, they would rest and sit on the boulders and the ground by the creek and take in the natural beauty of the area. You could sit for hours just watching the birds flying around and listening to their chatter. Occasionally a brush or wallaby would come close by on its way to get a drink at the creek. Usually, after a rest when lunch was finished they would pack up their picnic gear and head off home again to get back in time to milk the cows.

Tennis was one of their favourite sports and it was necessary to build their own tennis court. First, a site was cleared about 30 metres out in front of the house. Then with the dray and horse, they would collect old white ant nests and bring these back to the cleared area. Once these were offloaded they were crushed up with hammers, spread over the ground, watered and tamped down tight to form a hard surface. After a few weeks of this work, the tennis court was completed and they began to use it for play. After a while, a high fence was erected using bush poles and chicken mesh wire. Many afternoons would see them on the court playing tennis when the milking had been completed. On most weekends there was a crowd of people to play tennis. Esther became a strong tennis player as she played against the boys. Quite often beaten by the boys but this helped her in becoming a strong player. Clem and a neighbour, Baden Tonkin became very proficient at tennis and they competed in one of the Country Week teams in the Perth competition, working their way into the finals.

Many people came to the Kentish's at Bolinda Vale to play or watch tennis. Saturdays were very busy. After the morning milking and breakfast, the court was prepared for play. As soon

61

as two people were available, the games would start. As other people arrived the games would become more interesting with others providing different levels of ability. The play would continue but at mid-afternoon, the boys and Esther would be needed to do the afternoon milking. They would return to play immediately milking was completed. They probably worked a little faster. Most afternoons during summer were taken up playing tennis. With all this practice they all became good tennis players and community members.

The principal at North Dandalup School, Mr Anderson and his family were amongst those who competed at Bolinda Vale. His daughter Dorothy, now Mrs Charles Cowcher of Williams, well remembers these tennis games and comments on the great times they had at the farm. She also remembers Lance and Clem as being the two most eligible young men of the district.

Mr Anderson also had a block of land at Mandurah, which was not far from the beach. This was in a very pleasant setting with couch grass on the ground and paperback trees all around. It was, in fact, an ideal camping area and that is what it was used for. Ted and his family would take annual holidays and go camping at Mandurah, setting up tents on this land owned by Mr Anderson and going fishing and swimming. As they had no money to spare they purchased a used tent with fly. Then they cut light bush poles from the hill for the tent supports. The beds were constructed with light bush poles and old wheat bags cut to fit on for the base. The shorter poles were dug into the ground to form the legs. 6-foot poles were tied to these to make the side rails and 3-foot poles used for the cross members at each end. These would be carted down to Mandurah on the Ford "T" ute and then carted back again after their holidays, with the exception of the poles which were stored under a friend's house for the next years camp.

Usually, other people would be camping there at the same time. Hughie Manning was one of these and I can imagine that these young men would have got up to plenty of pranks to amuse themselves. Jack Philips was another to be at Mandurah at the same time.

Hughie Manning and his parents grew some potatoes so they took some of these to the camp and these formed their diet along with the crabs, whiting and other fish that they caught. Edie would make bread in the camp oven, so with bread and butter, along with the fish and spuds, they had sufficient food. Occasionally the boys would hitch a ride on Ward's fishing boat and help them with their nets. Upon returning, the boys were given some of the fish, so this also added to their food supply.

A few times, the junior members of the camping group had ice-cream eating competitions. However, there is no memory of who won or how much was eaten but you can form your own opinion on this. The ice creams were obtained from Ward's shop, which was situated near the beach at Mandurah. Mrs Ward would make the ice cream from fresh cream, custard powder with a few special ingredients and churn this mixture whilst it was kept cold by ice. That which was left over at the end of the day had to be discarded. We know who took care of this problem.

A crop of oats was grown each year to provide hay and chaff for the dairy cows and horses during the summer. Superphosphate was just coming into vogue at that time and this helped to improve crop and pasture production. The oat crop was cut for chaff using a horse (and later a tractor) drawn binder and the sheaves were put into stooks to allow the hay to dry and cure. The size of the stook depended upon the crop. The heavier the crop, the larger the stook, so that they didn't have to carry the sheaves too far. This harvesting was done towards the end of the year and quite often they would work into the night when the light was good enough. Once the sheaves were dry they were loaded by pitchfork onto a horse-drawn wagon and taken to a stack near the house. Some was put through the chaff cutter, bagged off into chaff bags and stored in a dry shed until required, while the rest stayed in the stack until the feed shed was getting empty then some more was put through the chaff cutter. This operation was very labour intensive and was done between milkings.

They noticed a small spring flowing from the side of the hill to the south-east of the house and they began digging down on it to form a dam for stock water. Lance and Clem moved a lot of soil with a scoop behind the horses until they encountered rock. They could see water coming from the rock so started breaking into the rock. Some help came from a Mr Franki who helped them with the use of explosives. After many attempts at blasting, cleaning and digging they got down as far as they were able to but the supply was not very satisfactory but quite suitable for stock use.

In suitable seasons, maize was grown and cut whilst green to make silage. The green material was cut by hand using special tools made by Lance from old chaff cutter blades. The cut material was loaded onto a dray then driven to the chaff cutter. The chaff cutter was positioned over the pit and the green material was fed into it and after cutting, the green material dropped directly into the pit where it was left to mature. After maturing, the silage was cut with a broad axe and loaded by pitchfork onto a dray and carted out to the dairy cows. The old silage pit was in the paddock just east of the present grain silos. The theory of silage was to improve the quality of the milk but after several seasons there appeared to be no improvement and this labour intensive operation was discontinued until many years later when machinery was available for the job.

Hughie and Cyril Manning spent many hours working with Ted and the boys in the harvesting of the maize for silage. Ted was surprised at the throughput that the four of them could handle. The ten-acre plot of maize usually took the four young men about a week to harvest and cut.

While this was going on, the property was being cleared so as to increase the amount of land that was available for grazing and cropping. From 1926 to 1945 this was done by hand and with the help of horses. In order to get the trees onto the ground for burning, first, they had to be dropped. This was accomplished in several ways. One way was to fell the tree using axes. As both young men were extremely fit and a bit competitive they would have races to see who would cut down the biggest tree the fastest. Having developed good axemanship the trees didn't stand a chance. Sometimes both of the boys or one of them and Ted (one person each end) would use the crosscut saw to fell the trees. This was the fastest way to get them onto the ground. This crosscut saw was also very useful in cutting up logs on the ground. Another way to fell the tree was with the use of a hand-operated tree puller. This was basically a large heavy-duty hand winch that was anchored to the base of one tree and the free end of the cable tied high up in the tree to be pulled down. Then the winch handle was operated until the tree fell. For some larger trees, several pulley blocks were necessary to improve the mechanical advantage so increasing the power of the tree puller.

The other way to cut down a tree was to arrange logs and other timber around the base of the tree and set fire to it. This was fairly common practice except for near roadways and fences or other places where the falling tree may damage something.

Once the tree was on the ground, any good timber was removed from them, and then the branches were cut off and with the help of the horses, arranged around the trunk and set on fire. Usually, the logs were positioned over the stump so it could be burnt out but some big trees could not be shifted until the fire had reduced their size. When these logs were on fire, they needed constant stoking to keep the fire burning hot enough to completely consume the green timber. This took a lot of time with a crowbar and horses with a drag chain.

Quite often the logs were too big for the horses to move them so they were cut with the crosscut saw or smaller logs were laid across the top of the bigger ones and set on fire. The fire would burn through the logs, dividing them, and then the horses could shift them for stacking and burning.

On the side of the hills, a short chain was attached to a stump and to the axle of the dray. With Jimmy the Clydesdale horse facing downhill and the lift created by attaching the short chain to the axle, most of the stumps could be removed with the boys and Ted digging around it and chopping off the roots.

In the years soon after their arrival at Bolinda Vale, Ted purchased a Clydesdale stallion from McLarty's of Pinjarra. He bred with this stallion and Jimmy was one of the progeny.

Stump bush regrowth was always a problem after clearing and this was the boy's job to control. This was done with the use of a mattock or with the horse and dray. For days on end, they go out after milking in the morning and spend all day grubbing out. Usually, if they were working close to home Esther and Edie would use the horse and buggy to take lunch out to Ted and the boys and boil the billy for a cup of tea. If the job were further away, they would take a packed lunch with them and boil their own billy. This is the way most of the land was cleared. It was very physical work and they would put in very hard and usually very long hours. While they were young and fit, this heavy work was not a problem but in later years, as with most men who had worked this hard whilst young, they suffered from many physical complaints, including arthritis.

These same methods were employed not only at Bolinda Vale as described above but also at the other properties that were purchased until about 1945 when they used a contract bulldozer for the first time.

Their first tractor was purchased in 1927 for about, £200/-/-. This was a steel-wheeled Fordson tractor with which ran on petrol for starting and warming up and then was switched to kerosene to keep it running. This tractor was used mainly for ploughing and harrowing. The belt pulley on the tractor was used to drive the saw bench for the cutting of firewood and also to drive a pump on the creek to pump water and probably the chaff cutter. Lance, being the mechanically minded person, was in charge of the tractor and other machinery. He drove the tractor for the most part and learnt the best ways to operate it to achieve the best results. At one time while he was still learning, he had a set of offset disc harrows attached to the back of the tractor. The tractor became bogged in the heavy soil. He wasn't able to unbog the tractor with the plough attached so he disconnected it from the tractor. Still, the tractor was stuck fast in the mud. He got some sleepers and attached these to the steel rear wheels and engaged first gear. As was fairly common with the machinery of the day, the clutch was fairly sudden and as he released the clutch the tractor reared up on its back wheels. Just before the point of it overturning and pinning him in the mud, Lance was able to depress the clutch and let the tractor's front wheels back onto the ground. Badly shaken he got off the tractor and sat down for quite some time to settle his nerves. He eventually got the tractor out of the bog and working again but he was always very conscious of the danger of the machinery.

The tractor was used for the ploughing now but the seed was still spread by hand. Harrows were pulled by the tractor after seeding. Super (about £ 3/10/-a ton) was applied by hand but whilst standing on the back of the horse-drawn dray. Ted would drive the horse and dray while one of the boys did the spreading. This practice continued until the late 1930s when a Horwood Bagshaw broadcaster was purchased. This broadcaster was attached to the rear of a horse-drawn dray and the hopper held one bag at a time and driven from the wheel. This then allowed them to cover more ground quicker, rather than work less.

Ploughing of the paddock was necessary to prepare a suitable seedbed for the seed and also to turn in any grass to provide some mulch. This was done in "lands" about 15 yards apart. A disc plough was drawn by the tractor and this would throw the soil to one side thus forming a ridge. So when a paddock was ploughed in this method, there would be a shallow depression and a low ridge every 15 yards across the paddock. Although this was the accepted method of the day it caused some problems. In heavy rains, these troughs would fill with water and then wash away thus causing erosion. One minor benefit was the retention of soil moisture but this was for only a few days longer than normal so was not of any real value.

Lance and Clem developed a good working relationship with each other as Clem was more attuned to the stockwork and Lance to the mechanical side of the farm. Both would work together on larger projects or with employees from time to time. Ted was maturing in age and was slowing down in his ability to work beside the boys but his organisational skills were still very strong and often clashed with the boys who had differing ideas on how to attack a problem. This regularly led to lengthy discussions after the evening meal.

Ted may not have been totally settled at Bolinda Vale, as he showed some interest in a property that was advertised for sale at Northam. This property would have been suitable for growing wheat and sheep, as he was more attuned to this type of farming, he investigated the possibility of purchasing the other property. He contacted the agent and went with him for an inspection of the property and in Ted's opinion, it would have been an ideal property for his purposes. However, although land values were very low during the depression, he would need to go into debt to purchase this property. The sale of Bolinda Vale would have been necessary to complete the deal and these same low land prices were working against him for this transaction. When he purchased Bolinda Vale, he had in mind to use it as a place to get established in the West and then find a more suitable property. He was loathed to get into debt at this time because of the insecure nature of the markets and after a lot of deliberation decided that they should stay where they were. Edie's health was markedly improved at Bolinda Vale and Ted was getting on in age and these may also have a bearing on his decision. They were settled in now and decided that the best choice would be to stay.

In about 1930 a dairy was built. As these were depression times and money was extremely tight, the building and other improvements had to cost basically nothing to enable them to improve their property. This type of cost-saving was carried on by the family for the rest of their lives, but they have been able to spend a little more freely in later years. Ted and the boys would cut poles from the bush for the posts and framework.

Second-hand iron was used for the roof and flattened out tar drums, from the recent sealing of Bunbury Road, for the sides. Bush poles were also used for the yards and bails. The two boys and Esther thought this was great to do the milking in a shed at last. It kept them dry and warm in the winter and shaded and cool in the summer. Although they probably didn't get a better return for the milk, the shed definitely made the job a lot more comfortable.

Apart from the dairy cows that they had on the farm, sheep formed another source of income for them. At one time a few were being killed in the paddock and they thought that foxes or dogs were the cause of this. One evening Ted and Esther heard a dog barking up on the hill so they took their rifles and went to investigate. Esther saw a dog moving around the paddock and got closer to investigate and saw it at a freshly killed sheep so she took careful aim and shot it. They buried it on the hill where it fell and next morning Ted advised Mr Wallace that his dog was found killing sheep and was shot. Mr Wallace refuted the claim that his dog would do that and went to retrieve him from the hill and buried it close to his home. Although this caused some ill will between the families for a short time they continued to remain good neighbours.

In 1932 a committee was formed in the Keysbrook district to build a community hall. The locals built the Keysbrook Community Hall after the committee raised money from donations and various fundraising functions. Timber was purchased from Whittackers and the State Timber Mill and the cost of the building didn't exceed £60/-/-. The main building was completed in 1935 and the floor was completed in 1937 when they were able to dance on a timber floor.

Table Tennis was a sport that was played frequently in the Keysbrook Hall. Members of the district took up many a night with this game. With the numbers of young people around there was a good bit of competition and their levels of skill were fairly high. Others came from North Dandalup to play and compete in the games that were very popular. One of the McCarthy girls and Esther were fairly evenly matched at the game and Esther went on to take out one of the trophies for her game.

The Keysbrook Hall was the venue for dances that were held fairly often. In later years at Christmas time, there was a Children's Christmas tree celebration put on with one of the locals doing the part of Father Christmas. In later years Mr Ivan Elliot and Lance played this part.

Another function that was held in that hall was the Australian Citizenship Ceremony. Several of the employees of the farm were Nationalised during these ceremonies.

About 1935 Mr Les Bee operated a Telephone service from the Post Office at Keysbrook. This was a very manual system and was prone to breakdowns. Occasionally those using the service would complain about eavesdroppers and at a suggestion of this, there was a clicking noise and the quality of the line would improve. Obviously, the eavesdropper would hang up at this time. After several complaints were lodged with the Telephone Service, both Clem and Lance requested that their telephones should be connected to the manual exchange at Serpentine. This was done in the early 1950s. The service was still fairly limited with it only being operated between the hours of 8 am to 6 p.m. during the week with 10 am to 12 p.m. on Sundays. The telephone wires were attached to poles alongside the road. Each pole was about 20 ft high above the ground with 8-10 cross arms attached. On each cross-arm had about 8 wires attached to it by insulators. This would allow for about 30 lines to be used at one time between Bunbury and Perth. In the 1970s this overhead system was replaced with the underground system that is used today. In the 1960s this manual telephone service was replaced with a fully automatic service that operated 24 hours per day.

A block of land near North Dandalup was on the market from a Mrs Dunlop. Although it was only 93 acres and covered in timber and scrub, it was a good buy. They purchased this land and set about clearing. There is a gravelly, heavy soiled ridge in the middle of the block and this was covered in heavy redgum trees and these were left behind after the scrub was removed. The lower section on the river flat was mostly tea tree and was grubbed out by hand. The north end was good soil and covered by redgum and scrub. Some of the largest trees are still standing today. A freshwater spring provides water for the stock. This block was fenced and young cattle or dry cows were pastured there. A young Arthur Gill was employed on the farm at about the time when clearing here was completed and there is an account of him droving a mob of cattle from this block to Bolinda Vale. Apparently one of the cows had recently calved and the calf was too small to travel far by itself. So Arthur carried it on his shoulders for about 7 miles while he walked behind the mob.

The Methodist Church sent Jim Morrell to Pinjarra as a Home Missionary. He was frequently at Bolinda Vale in this capacity and probably more often as a friend. Jim would assist in the Sunday Service and was a great friend to the family. While discussing his family history one Sunday evening, Edie discovered that Jim's adoptive mother, Maggie Graham, was one of her best friends while she was growing up in South Australia. Such a coincidence shows it's a small world. Jim was full of life and fun and one night when the boys were having a noisy discussion around the table one evening, Ted exclaimed, " Young men exhort to be sober-minded!" That quotation comes from the Book of Paul. At that point, Jim shut up like a book and appeared to be somewhat subdued. Ted had intended the remark in jest but it appeared that Jim might have taken it to heart. This did not, however, reduce his affection for the family.

Lance, Clem and Jim got on extremely well together and they got up to quite a few pranks. Because of the frequent visitors to Bolinda Vale, beds were set out along the verandas of the house. The girls and their friends were on the back verandah and the boys had their beds on the front verandah. The boys arranged a bell under the raised verandah beneath the girl's beds and ran a wire from this to a point near their beds at the front of the house. Not long after the girls and their friends got into bed one of the boys would give the wire a tug, causing the bell to ring. This startled the girls and they all looked around on the verandah and under their beds to find the source of the sound but could not. So they all settled down once again a little befuddled. About half an hour later the wire was pulled and the bell rang again. The startled girls all jumped out of bed and had another look for the source of the sound and found a bell under the verandah so they stuffed it full of socks and handkerchiefs. This quietened it down but the boys didn't realise that it didn't sound when the wire was pulled. Finally, the girls got off to sleep but then the socks and handkerchiefs fell out and the bell sounded again. The girls had had enough so they covered their ears and went to sleep. The boys got tired of pulling the wire and not getting a response so they got to sleep also. Next morning the girls told of their experience at the breakfast table and then the boys burst out laughing, it wasn't until then that the girls learnt of the wire and that the boys were pulling it.

The boys weren't the only ones to have the fun as Esther and her friends occasionally got one back on the boys. After making the boys beds they attached a string to the bedclothes of Jim's bed and ran it around the house to where their beds were. As it was dark he wasn't aware of its presence. The girls were listening intently and soon after he got into bed and pulled the rugs up, they pulled on the string and removed the bedclothes from him. A startled Jim recovered his sheet and rugs and found the string then realised that the girls had got one back on him.

Jim was well involved with the Christian Endeavour group in Pinjarra and the boys used to ride to Pinjarra frequently on their pushbikes. For night riding, carbide lights were attached to the front of the bikes to illuminate the road. A truck would bring down a group of people from Perth and they would join in Fellowship and services. Lance became involved with one of the young ladies who were in this group. One night, after Fellowship he closely followed the truck that they were riding on and grabbed hold of the back rail and "hitched a ride". While he was hitching this ride he furthered his discussions with this young lady who was working in a grocery shop in Armadale and later in Subiaco.

Vera Lavis lived as a young girl and teenager around Woodanilling, which is a small farming town near Katanning in the Great Southern region of Western Australia. She was from an English migrant farming family who had suffered hard times, particularly in the early years of the depression. As a teenager, she took a job with Mr J C Gomm, Pop as he was to become affectionately known, who was a carpenter employed by Richardson's in Katanning. The business supplied hardware and farm supplies to the people of the district. Later he began his own business when he purchased a grocery store. His son Les was a Baptist minister in Katanning at the time but he moved to Fremantle. Soon after this, Pop sold his business and moved with his wife and Vera to Armadale and set up shop there. Lance visited her frequently at Armadale when he would ride in the 19 miles on his pushbike. Later he travelled into Subiaco and she travelled to Pinjarra for Christian Endeavour at every possible chance and frequently visited the farm and had fellowship with Lance's family. Pop was on the move again when he bought and operated a shop in Subiaco. They had living quarters at the back of the shop. Vera shifted with Pop and his wife, who was very ill at this time, and they operated the vegetable business and Vera nursed his wife as well as assisting Pop.

He would frequently deliver groceries and vegetables to the construction gangs at Canning Dam. This caused him to use the road down the Roleystone hill where he had brake trouble at one time. Vera's sister Molly was also living with the Gomms at their shop in Subiaco at that time.

On March the Thirteenth of 1937 Lance and Vera became married. The ceremony was performed in the Leederville Baptist Church. Enid relates a small crisis just prior to the wedding. While Vera was getting dressed, she caught the hem of her wedding gown and tore it. They were a little late in getting to the Church as a repair was done to the dress.

Vera's Bridesmaid was Connie, her sister, with Esther being the second Bridesmaid. Keith Nicholls, Lance's cousin, was Groomsman with Clem being the Best Man.

After the reception, Lance and Vera borrowed Enid's little Austin Seven car and went to Bunbury for their honeymoon after spending their first night at Ravenswood Hotel.

A new house was built for Lance and Vera. They affectionately called this "The Wee Hoose " and is situated near the creek to the northwest of the old home on Bolinda Vale. This was a 2-bedroom house with a kitchen and lounge room with a verandah front and back. The front verandah which faces the west had canvas roll-up awnings fitted so that they could block out the sun and heat in the summer. After sunset, the awnings were rolled up to allow the breeze to cool the house. Behind the house was the washhouse and further down the back was the " Thunderbox " lavatory. The house was built during the height of the depression and the job was done on a very tight budget but the timber-framed, weatherboard and the corrugated iron house was very respectable. Later a car shed was added just away from the house on the north side. Pop Gomm came to live with Lance and Vera later and at that time built some additional rooms to the house.

With the added benefit of an extra pair of hands, Ted was beginning to ease off a bit as he was getting on to 72 years of age. Vera was very capable on the farm and was helping with the milking and fruit picking, in particular, always whistling while she worked. Much of the riper fruit was bottled and preserved for later use and Vera was very adept at this work and always able to find ways of doing things that were cost-effective and money-saving. She had a vegetable garden close to the back of the house so she was able to keep their living expenses very low.

Vera was kept busy with the farm work and also with keeping up with housework. On April the 5th 1938, Vera was admitted to Pinjarra Hospital where their first daughter, Gwen, arrived. Upon returning home Vera had an added workload but Esther, Enid and their mother would help wherever they could.

Hand milking continued until about 1940 when a MacDonald Imperial Milking Machine was installed. This was powered by a Gray stationary petrol engine (Graham has recently restored this). This enabled them to milk a larger number of cows and increase their production. There was an employee at that time by the name of Stan Chamberlain. He said to the Milking Machine salesman " Don't sell Mr Kentish a milking machine as then we'll work faster and it's the only chance we get to sit down". At about this time they also began supplying "whole milk" to the Perth market. The return for "Whole Milk" was a little greater than was previously received but with the added bonus of a quota that allowed them to supply a larger amount of milk.

The Milk Board Of Western Australia required them to build a new dairy prior to the supply of " Whole Milk ". With some skilful arguing they were able to convince the Milk Board that even with a new dairy, the quality of the milk that they would produce would not be improved. They continued milking in this old shed until they completed the construction of the new dairy in about 1946.

Vera had a vegetable garden near the back of their house and she grew some very good strawberries. She noticed that she wasn't picking very many at one time and Pop told her that he had seen little Gwen picking and eating them. Vera told Gwen that she was not to touch the strawberries. For the next few days, Vera picked quite a few strawberries and was happy again that everything was all right. Then the numbers fell off again. She just happened to be looking out the back over the strawberries and she saw little Gwen on her hands and knees bending over, eating the strawberries straight off the plant. "But mum, I'm not touching the strawberries" was her answer when Vera challenged her about it.

A neighbour, Mr Les Robinson, was ill during one years' hay time and had to spend time away from the farm. One moonlight night Lance, Clem, Esther and a few other locals went to the Robinson farm and stooked all of the hay which had been cut and laying on the ground. Les was worried, when in hospital, that all his hay would be ruined but was delighted and surprised upon his return to see stooks of hay in his paddock.

The family continued with their Christian activities and the word had spread about these Church Services being held in the family home. A band of Salvation Army lads would travel out from Perth each Sunday and participate in the Service and stay on afterwards for a cup of tea and probably a meal. Esther received a letter from one of these lads, an Allan Uren, who requested that she might correspond with him. This was to prove to be a very lasting relationship.

Allan Uren was living in Maylands where his parents had a small shop. Allan had Salvation Army connections and would attend church with them for the fellowship. They became engaged to be married in 1940. Allan's Salvation Army friends were about to join the Armed Forces and Allan decided that he also would make the commitment to serve his country. They joined the 2-16 Infantry as Army Bandsmen. This suited Allan, as he would then be with a group of Christian men within the Army. About this time Ted had a deep discussion with Esther and suggested that they should become married prior to Allan going away to war. He felt that it would be easier for Esther, should Allan not return, to go through life as a married woman than a spinster. Ted's sister, Alice had lost her fiancé, during WW1 and she had a very difficult time and Ted could not let this happen to Esther. This had a further advantage, in as much as Esther was able to see much more of Allan prior to his going away than had they not married before, in camps etc.

On June the Eighth, 1940 they had their wedding ceremony in the Anglican Church near the Serpentine River Bridge. The Rev Ron Limb, who was stationed at Pinjarra at the time, and Mr Doley, who was in the Ministry at Booleroo Whim but now retired and living near North Dandalup, officiated at the ceremony with Allan's brother being the Best Man, Enid acted as Esther's Bridesmaid and little Gwen (Lance and Vera's daughter of 2 years & 2 months) was her Flower Girl. Esther wore a lovely white wedding gown that had been made especially for her. Her father was very emotional (a common family trait) and just before the parade down the aisle; Esther needed to settle him down rather than the other way around. Ted was 75 years of age at this time.

Esther and the others had been teaching Sunday school at Serpentine, riding their pushbikes each Sunday. These Sunday school children provided a lovely " Guard of Honour " for the Bride and Groom outside the Church after the ceremony.

A grand reception was held at Bolinda Vale on the back verandah for the newly married couple. Edie, Enid and Vera provided the refreshments for the wedding party and the guests.

Allan and Esther went to Kalamunda for a one week's honeymoon with £10/-/- to cover costs.

Allan went away to Army camp at Northam a week after the wedding and Esther returned to the farm. She was able to visit him at Northam on several occasions. Allan shipped out in October 1940 and by Christmas, he was in Palestine. He saw no action here but was doing a lot of training and some sightseeing. Allan did a lot of writing home and Esther has kept the letters of this era and they would be very interesting reading.

The 2/16 Infantry Battalion was shipped out from the Middle East and taken to the fronts in New Guinea, via Java and Fremantle. Esther was able to spend some brief time with Allan while his boat was in Fremantle.

Allan spent much time tending the wounded until he himself was injured by the Japanese. Esther received a telegram from the Army to say that he had been wounded in action and to express their deepest sympathy. The telegram arrived at the Keysbrook Post Office that was run by Mrs Bee. She gave the telegram to Clem who was there as the mail arrived; he took it home and gave it to Esther when she returned. A few days later she received a letter from Allan explaining what had happened. He had copped a bullet wound through the arm and hand. After confinement in New Guinea, he was transferred to Hospital in Queensland.

As the trains were reserved for troop transport in 1942, Esther travelled to Sydney onboard the ship "Katoomba" to visit Allan in hospital. He was discharged from the Army in 1943 as he had returned to Perth and was working for Mr Bell at his hatchery. Allan had sold his Motor Bike to Clem before he went away with the Army so now he bought it back again and this was their transport. Esther and Allan were building their home at Wyatt Road in Bayswater and would visit the farm frequently but with petrol rationing this became difficult.

They have brought up their family and still live in the same house although it has improved over the years. Chicken sexing was one of Allan's occupations and he spent some time growing gladioli for the flower market. His rotary hoe was usually busy with Allan at the controls, hoeing gardens and firebreaks. Graeme was to carry on with this in later years. Allan and Esther are ardent travellers and have their van set up to take off at a moment's notice. They frequently visit Bolinda Vale and help Colin with his relearning.

During World War Two there was a substantial Army Camp just to the northeast of Mundijong, George Marriott and Roy Lane were two of those troops who were stationed there for a time. Les Gomm, the son of Pop Gomm, was the Padre at this Mundijong Army Camp. So with this affiliation through the Gomms some of the soldiers would be taken to the farm for visits and Fellowship. One of these frequent visits was " singsong " night on Wednesdays. George formed a lasting friendship with the family with many reciprocal visits over the years only ending with his death a few years ago in 1990. Les Gomm went on to become the Minister in the Adelaide Baptist Church in the 1960s.

Fred Burt was the son of the Methodist minister from Pinjarra, who was having health problems and came to stay at Bolinda Vale for some time. He survived on a diet of oranges and he maintains that this is what corrected his medical problem. He went on to become a Church of England Minister. As fate would have it, he was Padre for the 2/16 Infantry Battalion and met up with Allan again whilst in Palestine.

Soon after War was declared both Lance and Clem attempted to join the Armed forces. Clem tried on several occasions to join the R.A.A.F. but his application was rejected each time on the basis that he was involved in "essentials services" (that of providing materials and foodstuffs for the war effort and the population). Lance filed several applications to join the Armed Forces but each of these was also rejected for the same reason. So due to the Manpower Act, they were deemed to be of more benefit to the Country to stay on the land and provide this service. Their father was getting on in years and not very well and if the boys were not there to work on the farm, it would have gotten into disrepair or even disposed of. Although disappointed at not being able to be part of the "big action" Lance and Clem accepted the situation and continued with the farm.

Reliable farm staff was a problem to find prior to the war years but now with many men going away, this problem was much worse. This caused everyone on the farm to work even harder and longer hours or even trim their activities to work around the staff that was available. Holidays were curtailed during this period and with some mechanisation, they were able to cope with the situation. Alice Hitch was one of the women staff that was employed at the farm during this time.

Some farm machinery was available now so Clem and Lance got hold of a hay baler that was driven from the belt pulley of the tractor. This was a stationary unit and the hay needed to be carted to the baler. The material was forked into the chamber and the plunger would compress it. When the bale was long enough a board was put in to separate it from the next bale and they tied off the string by hand. Now they could make about 300 bales per day. This now made the handling of hay much quicker and more convenient. The stacking was a breeze with these rectangular bales. There was still a fair amount of manual work involved but this new hay baler was a forerunner to increased farm mechanisation.

Lance and Vera's second daughter, Marjory, was born in Pinjarra Hospital on 26th of April 1941. Soon after their return to the farm, Pop Gomm added a few rooms to the north side of the "Wee Hoose" where he would make his home. One of these rooms had a seat built-in under a window. The girls were often found playing in and around it. Vera had nursed Pop's wife during her last days and he found this way to repay some of the kindness shown him. He was a great friend to all and helped the family in the home and around the farm, thus alleviating some of Vera's workload.

Kentish Bros

Ted was an old man. He had worked extremely hard and for many long hours over his lifetime. He had seen many changes to society, properties and family during his lifetime. The Church and family were the things which he held as the dearest to him and these were with him even after the end of his life. Edward Joel Kentish passed from this life on a quiet sunny afternoon at his home, Bolinda Vale, on June the 26th 1942 at the age of 75 years and 3 months.

In his Will, he bequeathed the property of Bolinda Vale to his wife Edith. She could foresee problems with this arrangement and after a family discussion, decided to divide that portion of the property into equal shares for her 4 children. The property was professionally valued and divided. While all agreed that this would probably be a better arrangement, Lance and Clem could also envisage further problems. As that arrangement stood, they would be developing the land to improve its production and the girls would benefit from this by way of an improved asset for no input. While they did not wish to be selfish they were not happy to do this. At this time Lance and Clem raised sufficient capital and paid out the two girls their share of the property. This allowed them to work the land and build up their property while allowing the girls the capital to carry on their own lives. Esther was now married and had a breadwinner and this capital enabled them to settle into Wyatt Road and build their own home. Enid was being courted by a young gentleman who played music and things were looking serious so she could foresee a good use for her capital. This arrangement between Lance and Clem was the beginning of "Kentish Bros. ".

Enid was teaching music at Pinjarra and occasionally on her return journey, she would be met by a young chap driving a Ford truck fitted with a gas producer. Enid would accompany him and spend some time at his poultry sheds and a "cuppa" in the incubator room. Reg Ingpen was his name and he often came to Bolinda Vale to visit Enid and play music. Reg was a very gifted musician having learnt, as a lad, the flute. He was adept at playing the piano and violin much to the delight of Enid, her mother and everyone at Bolinda Vale. Reg earned his living operating his poultry farm and selling the skins of the rabbits he had trapped in the hills.

Radio was in its infancy at this time and Reg was very adept at making and repairing radio apparatus. He had made several "Crystal Sets" that he would listen to each night. Reg and Enid corresponded by letter frequently and Enid was taken by the nature of the man who wrote these letters. Reg proposed to Enid and on July 28, 1943, they become engaged to be married.

Their courtship continued to flourish and they became married on November the fourth, 1944. The ceremony was held at St. Steven's' Anglican Church on the bank of Serpentine River. Enid wore a long white frock with a veil borrowed from Mary Lunn. Esther and Dorothy Day, who were dressed in blue frocks attended Enid while young Gwen and Marjory (Lance and Vera's two daughters) were the pink dressed flower girls. The reception was held in the Serpentine hall with Mrs Perrett and Mrs Manning supplying and arranging the flowers. Enid's mother made her famous yeast buns and Mrs Pennell made and decorated the wedding cake. After the reception, Reg and Enid retired to Allan and Esther's home in Wyatt Road Bayswater. Among their wedding gifts were petrol ration tickets (wartime restrictions) and clothing tickets.

They made their first home on "Hilcot" where they lived until their house was built on "Ellora". Enid and Reg spent some memorable time planning the house and surrounding buildings. The house which cost £52/10/0 to build was constructed by Mr Philips. Mr Harry Fawcett built the chimneys for £25/0 /0.

Although the house was not lined they moved in on 19th January 1945 and attended to the lining at a later stage. The building which was the incubator room at Hilcot was moved to become the wash house at Ellora. Reg had a well dug and fitted it with a windmill, pump and tank. Enid and Reg worked together running the poultry operation with the collecting of eggs, cleaning and packing of them, cutting and chaffing grass, using an old hand-operated chaff cutter.

Enid and Reg still resided in the same house until just recently. They have raised their children and seen them grow up and move away except for Robert who has stayed on the farm with his wife in another house. Both Edward and Joan now live in South Australia just out of Adelaide. Enid has continued with her teaching and has over 300 pupils that she has given music instruction to, over a period of 65 years.

Gwen began her schooling when she attended Keysbrook State School in 1944. There was daylight saving during the war years and School didn't start until 10.00 am, much to the dismay of the locals and parents. Most farm children helped around the farm when they weren't attending school and with daylight saving there weren't enough daylight hours when the children returned home from school in the afternoon but it allowed them more time in the mornings. Gwen was able to assist her mother around the house and help with her little sister, Marjory when she wasn't at school.

Vera attended Pinjarra Hospital once again for the birth of a third daughter who they named Lorna. She arrived on July the 12th 1944. Vera's sister Molly came to stay at the "Wee Hoose" to look after Lance, Gwen and Marjory while Vera was in the hospital. She brought with her Barry and Laurie, her two sons who were about the same ages as Gwen and Marjory. When Molly called them for lunch one day she could not get any response, so she continued calling. Still getting no response for some time she began to get worried and started looking for them. She eventually found them up to their knees in pig muck in the pigsties. She took them home and washed them down with the garden hose before she could put them into the bathtub to get them clean. Even after a good scrub they still smelled of the pigsty for several days.

Clem decided at an early age that he would participate in local government affairs and was elected to the Serpentine - Jarrahdale Roads Board in 1944. He was Chairman for 1944-1945. He held his position on the Road Board until 1957 when he resigned.

A property about one mile to the north along the Bunbury Road came on the market and after inspecting it, both Lance and Clem agreed that it would be a good purchase. They bought this property, known as "Mount View" from a Mr Wells. When they took possession of it they cleared away all of the rubbish lying around and found a lot of useful (to someone) implements and farm items and tools. They held a clearing sale to dispose of the unwanted gear and deposited a fair amount of money to their loan account. The sheep which were on the property were sent off to the abattoir as some were showing signs of footrot. The old Mount View homestead was situated near the railway line down Fisher Road. This house was used for staff accommodation for many years. Near this house was a set of sheep yards and a shearing shed complete with a well and windmill. Between the house and the railway line, there was an old orchard and closer to the house were some large grapevines. These were the Musket variety and a cutting was taken from here and planted at the new house when it was built.

The old shed that they had been using for milking the cows on Bolinda Vale was in need of extensive repair and becoming too small. Rather than spend time and money fixing it they decided to build a new dairy.

To construct the new dairy they used gravel and sand which came from the creek, mixed with cement to form the concrete which was used for the floors and walls of the dairy. All materials were loaded onto the horse-drawn dray by hand using shovels and then mixed by hand on-site. They were able to employ some local men to help with the construction. Pop Gomm was the overseer of the building, Mr Harry Fawcett (from Scarp Rd.), Mr Maurie Fisher (from across the railway line) were some of those who assisted in this work. After the concrete floor was laid, timber plank formwork was erected to form the concrete walls. Timber trusses were fitted on top of the walls and asbestos roofing material fitted. The Mac Donald milking machine was installed and this was one of the newest and most modern dairies in Western Australia.

Soon after the new dairy had been constructed, Lance and Clem were looking for an alternate supply of water. Mr Westcott who had a property west of Keysbrook had some success at divining for water and suggested the most likely place to dig, just happened to be in front of the dairy. The young men armed themselves with a pick, shovel, windlass and bucket then began digging. They continued with this very difficult and dangerous work until they reached about 60 feet. Although they encountered a patch of water washed stones they didn't find any water. Mr Westcott couldn't understand this so he pitched in to help with the job. They reached 75 feet and only a trickle of water on top of a hard base so they had to stop at that. A very frustrating result for all of the hard work that went into the job. However, they put a windmill to work and used all of the water that the well had produced. A Tank was fitted to the top of a stand by the windmill and this provided some water for the dairy and the stock, complementing the water supply from the creek.

Marjory was only a toddler at the time when she had a large doll. She was playing around near the dairy and stables and the well that had not long been completed. She was getting upset with the doll because it would not talk to her. Eventually, she had had enough of no response from the doll so she walked over to the well and threw it down. When she realised what she had done, she got upset and then the tears started to flow. Someone saw the situation and went down the well and recovered the doll. Although it was wet and soiled Marjory cuddled the doll and the tears went away. I don't think the doll would speak to her now even if it could.

Their father had been a tremendous guiding influence to Clem and Lance but many of his farming ideas were outdated. After his passing, they were able to operate the farm in the manner which they considered to be the best way. They made many decisions regarding the methods of operating the farm and of carrying on the business, all of which have proved to be successful. Lance and Vera had begun their family so Clem and Lance decided that they would work their way to building up a farm that could easily be divided to make two separate units. This would give each of them a farm in their own right that each could operate with their own families. This proved to be very good forethought but was many years before it eventuated.

Now that the farm was providing a steady income from the production of Whole Milk, Lance and Clem were able to extend themselves and purchase more land around their farm. Much of this land was bush or fairly roughly cleared so they were still working hard to develop this and bring it into production. The dairy herd was increasing in size and performance as their landholding increased and property improved.

Mr Gobby had a property that he had purchased from Mr Mathews which adjoined Bolinda Vale to the south. He was running sheep and during the good seasons did very well out of them. However, he could not understand what was wrong with his sheep at one time so he called on Clem to determine the cause of the problem. The sheep were flyblown and when Mr Gobby saw the damage to the sheep and the agony that they were in, decided that he didn't want anything more to do with sheep and decided to accept Lance and Clems' offer of purchase. This was 160 acres and being only divided by a fence line, was an ideal addition to their farm.

Vermin are always a problem with farming and during the 1940s there was a plague of rabbits. "1080" poison baits would be laid out around the paddocks and in the next few weeks, the dead rabbits would be picked up. Quite often Gwen would be driving the horse, Lilly, and cart while two men would walk alongside and throw these dead rabbits onto the cart. They would do a few loads each day so that would represent about 2 tons of dead rabbits daily. Lance has not been able to eat a rabbit stew since.

The class size at Keysbrook School was fairly constant and Marjory attended for her first year in 1947. Gwen was with her so she was able to "show her the ropes" during her first few days. About 10 or 12 pupils attended with one teacher who had control of all grades. Her schooling was interrupted as she became ill with Rheumatic Fever.

Another cottage was built on the farm, this time to the east of the main home. Some of the first people to reside there were the Marsdens. Roy and Ken worked on the farm and their sister Margaret helped Edie, then Vera and Ira in their respective homes. They stayed on for a long time and became very good friends as well as employees. Margaret stayed in the district and married George Elliott. Ken married Helen Pollard from Keysbrook who worked for Lance and Vera on Mount View at one time.

Pop Gomm owned an old Chevrolet van which he used to deliver vegetables when he had his business. He would attend church with the family and occasionally take the children in his Chev van. Gwen remembers one day when they were returning from church in Jarrahdale when whilst coming down the hill they watched a wheel going past them. Pop stopped the vehicle to discover that it was his wheel that had come off. Vera relates a story when Pop was delivering vegetables in the Roleystone area. He was returning from his deliveries and travelling down the hill when the brakes of the old Chev van failed. Fortunately, the traffic was light in those days and there was no accident. He was able to slow down the vehicle by using the gears and eventually the hand brake. When the van came to a stop he sat there shaking. It would have been a frightful experience in the old van on the bone-shaking road, probably at a fairly high speed.

Miss Connie Davis was teaching at Keysbrook at about this time and she was boarding at Bolinda Vale. Esther was now living in Bayswater, Lance and Vera had their own home on the farm, Enid and Reg had their home at "Ellora" and Clem was living at the family home at Bolinda Vale with his mother. Connie had been staying there for some time and fitted in very well with the family when Edie had to go to the hospital for a few days with a chest infection. Clem and Connie were sharing the house alone and everything was going well and above board but Enid got a bee in her bonnet and told them that this was an "improprietous situation" and that the arrangement should stop. Upon hearing this comment, Connie told Enid " Go and jump in the lake".

At the completion of the milking, the cans of milk were loaded onto a wagon and taken to a ramp which was situated on the roadside near the front gate. From here it was collected by the cartage contractor and delivered to Cooksley and Dreyer who had a milk factory in Claremont. This arrangement continued for many years. Masters eventually bought out the business of Cooksley and Dreyer and later the milk truck would collect the cans direct from the dairy.

1947 was the year that mechanisation took a leap forward. Clem and Lance had attended a Field Day at Cardup to look at a new system for baling hay. Although Clem bought a new Vauxhall car instead of a baler at this time they both decided that this new system is what was necessary for their farm. Later in the year, they purchased a new International W4 tractor and the new B series McCormick pick up hay baler. These two went together well as a Power Take Off Shaft transmitted the power from the tractor to the baler. This enabled the tractor and baler to travel around the paddock and pick up the hay direct from the ground. The old Fordson steel-wheeled tractor had the hay rake attached to it and this preceded the baler around the paddock forming the cut grass into a windrow for the baler to pick up. A sledge which Lance had made was attached to the rear of the baler and a man caught the bales from the rear of the baler and made a stack of about 20 bales in this sledge. When this stack was built it was pushed off the back of the sledge with the help of a crowbar.

Now when the hay was baled there were these small stacks scattered around the paddock. A wagon and a few men were necessary to throw these bales onto the wagon and cart them into a central stack. This part was still done by hand. Later on, a hay loader was attached to the front corner of the wagon and this picked the bales up from the ground and elevated them onto the wagon where one man could easily stack them onto the wagon. When this happened the bales were left on the ground in a single line behind the baler where the loader could collect them from. A motorised elevator was used at the stack when one man on the wagon would take the bales from the stack on the wagon and feed them onto the elevator. Another man would take these from the top of the elevator and form a large haystack. This system was widely used until the late 1970s.

Early in 1949, Vera was confined in Pinjarra Hospital for the birth of her fourth child. On January 27, a son, who was named after his Great Grandfather, David Joel, was born. He was the first son of the new generation and would carry the Kentish name on for further generations.

The School at Keysbrook was graced with the presence of yet another Kentish as Lorna began her schooling in 1950. Marjory had been there for 3 years and was able to help Lorna adjust to the change of activity of the school.

Clem was the last of the four children of Ted and Edie to get married and during 1950 he attended a dance and birthday party that was given for one of his acquaintances in Perth. He had a very pleasant time and danced with quite a few of the young ladies who were present at the dance. One of these young ladies was an Ira Forbes Smith who was a very talented artist and designer. She was the daughter of a doctor who was living and practising in Subiaco opposite the Kensington Hospital. His surgery was at the side of his home. Dr Eric Smith had studied at the University of Edinburgh and also in Los Angeles.

Ira had studied Design and Commercial Art from 1941 - 1943 and General Art 1943 - 1945 at the Perth Technical School and graduated with a State Art Teaching Certificate. She had developed a personalised style in both design and general art. Her first exhibited works were a design for needlework " Roses & Violets" that was shown when she was aged nineteen years at the 1939 Annual Exhibition of the Society of Women Painters. By 1941 she had developed her own style that was different from the "English" style that was prevalent at the time. In 1947 she exhibited some of her oil paintings in conjunction with several other artists.

Clem would visit Ira at her home from time to time and they had a lot in common. Ira was also a good cook and Clem thought this was great. He also had a lot of respect for her ability with art designs and paintings. Her father was a little hostile with the idea of a marriage, as Ira was the mainstay in her home. Due to her mother's illness, Ira was needed to run the house.

After the afternoon milking, Clem would drive his car to see Ira at Subiaco. Sometimes he was quite tired after a days work and then on the way home one night he was stopped by the police late at night and cautioned for not using his indicator when turning. Ira visited the farm frequently and Marjory remembers her sitting in the car as she arrived wearing a full skirt. She had it tucked up and over the seat so as not to have it creased.

After their courtship and engagement, Clem and Ira had their wedding in Wesley Church in Perth with the reception being held in a nearby restaurant. Baden Tonkin of Keysbrook, Clem's old tennis mate, was the best man with George Quick, Clem's second cousin (Lorna Dick's brother), was the groomsman. Ira's long-time friend, Meline Guilespi (now Luke) was her bridesmaid with Ira's sister Valma assisting her as the second bridesmaid.

The odd one or two of Clem's friends were pranksters and Clem had just had his car done up, so he hid it in a local garage in Wellington Street so that they couldn't get to it. Some of his friends were a little disappointed that they couldn't dress up the going-away car.

For their honeymoon, they travelled to Geraldton and stayed in the Victoria Hotel for a week or so and spent many hours swatting mosquitoes during the night. They returned to Ira's home with the intention to stay for a few days more but on their arrival, they were greeted with a message to return home urgently as Lance was having some difficulties at the farm.

Mr Bob Stidwell did some work on the old house with relining and repairs. So that Clem and Ira could have the home to themselves, Clem and Lance had Bob construct a Granny Flat next to Esther and Allan's house in Bayswater. This pleased Esther no end as now she could spend more time with her mother. The flat had all the necessary equipment for Edie to care for herself and did not increase Esther's workload with her family initially. However, in the late 1950's she had a fall and got to the stage where she wasn't able to look after herself effectively and the family organised for her to go into a home for the aged. After trying several homes they found one that was most suitable in Mount Henry. She was resident at the Mt. Henry Hospital until, with failing health, she passed away in January 1960 at the age of nearly 83. She was buried in the Serpentine Cemetery in the adjoining grave to her husband Ted, who had been there since 1942.

The daughters of Lance and Vera were quite investigative. They wanted to know everything. Lorna remembers them throwing the old cat into the creek to see if it would swim. Much to their amazement, it could swim very well. This also answered another question that they had. This old tomcat would come home after a night out in the bush. Most times he would return with a rabbit in its mouth. He would go hunting up in the hill and sometimes they had seen him coming home. He would need to cross the creek to return home but until now they had not seen this. So their little experiment solved several problems.

1951 saw Vera confined in Pinjarra Hospital to give birth again and this time she had another son whom they named Neil. He would also carry the Kentish name into the next generation.

Horses were still being used on the farm for many jobs and Gwen and Marjory can relate memories of them. Most of these were Clydesdales. Bonnie, Jimmy and Lilly were draught horse. Blossom died in the early 1950s after she got into the wheat for the chooks. She ate too much, became bloated and died. The girls were very upset one morning when they looked out into the paddock and saw the old horse laying there all bloated up with her four legs sticking straight out.

The girls were a big help around the farm between schooling. Quite often during fruit picking season, you could see them with a horse harnessed to a cart, standing on the side of it picking the fruit from the trees. If Lilly was being used she occasionally would just gently move forward, grasp an orange in her mouth, squeeze it to break the skin and swallow the juice. Then spit out the skin. Jimmy and Bonny were not so gentle and occasionally caused a few problems.

Gwen was breaking her neck to leave school to come home to work on the farm. She stayed at school until the day before her 14th birthday, which occurred on a Saturday, and stayed to work on the farm until the month before she becomes married in June 1957. Gwen was very often riding a white brumby called Trixi. He was a powerful-looking horse and worked well with cattle. It was about 13 years since Esther had left the farm and it the men reckoned it was good to have an energetic young lass around again. There wasn't any work on the farm that Gwen wouldn't tackle. Gwen's main job on the farm was to feed the calves near the dairy. The calves were taken from their mothers after 2 or 3 days and kept in paddocks, according to size, and were hand-fed twice daily. The males were castrated and kept until they would return a good price for beef and most of the heifers were kept for replacement dairy stock. Lance had made a trolley about 4 feet square and this was loaded up with about 4 cans of milk and this was dragged over the calf yards at feeding time.

Another of Gwen's main jobs was to take care of the orchard. The men had constructed a diversion in the creek so they could put a sheet of iron across this structure and divert some of the water to channels which had been dug through the orchard. This would provide irrigation for the orchard at a very affordable price. The old Fordson kerosene tractor was kept near this area of the creek and was driving a large pump with its belt pulley. This pump would deliver water to a tank up the hill and gravity would give a good pressure at the dairy and houses. A 4-inch asbestos cement pipeline was laid from the tank to the dairy and generally, the flow of water was good.

Lance and Clem, with the help of Pop Gomm, built a new shed behind the very old shed that was used for the first dairy. Bush poles were used for the framework, used corrugated iron was fitted to the walls and they purchased new iron for the roof. In this, they fitted a new fruit grader for the faster and more accurate grading of the fruit. The shed was divided into two sections with one section having a raised floor where the fruit was stored and the packing boxes were put together. The timber for these arrived pre-cut and they were assemble over a jig and then nailed together. In the ground-level section, the fruit grader was situated along with the workshop and forge. They had set up an engine outside the shed and a long layshaft to near the tops of the shed post inside the shed. The engine was coupled to the layshaft with a flat belt and pulleys. To the layshaft in the shed was fitted several pulleys which provided power for several items. These included the fruit grader, post drill, double wheel grinder, lathe, and several other items. The forge was set up near the front opening of the shed with the large anvil next to it. A hand-operated blower fed the air to the forge and in later years, Lance's boys were to spend a few hours swinging on this handle to help their dad. Lance did a lot of maintenance of equipment in this workshop.

Inexperience is one of the main causes of farm-related accidents and Gwen was involved in one of these. She needed to get onto the roof of the dairy to check out a leak of the feed shed that was bothering her. She set the ladder up to the gutter and proceeded to climb. When she was about halfway up the ladder toppled over and Gwen was unceremoniously dumped onto the ground. Not only was her pride hurt but also her backside. Someone came to her aid and she stood up fairly soon after but she left the roof work to someone else.

In this feed, shed was kept the milled grain and feed supplements for the dairy cows. Apart from crushed oats in bags and bags of pollard, there were drums of molasses. The kids would have a great time mixing some pollard with a bit of molasses, or the other way around, and eating it.

Pop Gomm's blue Chevrolet Van had some slogans painted onto the front and rear of the roof. One had "God is Love" and the other had "The Gift Of God Is Eternal Life". Gwen refers to a time when she had gone with Pop Gomm to Church in Jarrahdale. After the service, they continued on to have a cuppa with the Jubbs who lived close to Jarrahdale. Barry and his sister Verna were discussing with Gwen about spreading the word of God. One of them jokingly said, " Put a bomb under it, that'll spread the word of God ".

In the year that Queen Elizabeth ll. was coronated, 1952, Marjory and Gwen went to Perth with Pop Gomm in his Chev. The Causeway had a plank decking in those days and the surface was very rough. The wooden wheels of the Chev would rattle badly on this road.

In the early 1950s Pop Gomm built himself a house at Busselton. It was a fairly modest asbestos covered house but it was his pride and joy. When the house was completed he moved from the farm to live in his new home. Lance and Vera and family would visit him regularly and would take their holidays in Busselton. There was a large camping area set aside just to the east of the town adjacent to the beach. Lance would set up the tents for the children to sleep in and also one for them to use as a living room. Most years they would spend a fortnight at this camping area and enjoy a good relaxing holiday away from the pressures of the dairy and farm. Swimming was a favourite pastime but there were some stingers in the water and this made it unpleasant sometimes.

Imagination is something that we all have and use. When David was about three or four years of age, he didn't have anyone his age around so he used his imagination and invented a friend. He was always blaming "Rosie" whenever he did something wrong and would talk to his imaginary friend frequently.

Ira was having some difficulties with her health, particularly in the springtime with her hay fever. She suffered badly with this complaint and it was with her for most of her life making things difficult for her. She was also suffering from a diabetic condition for which she required daily injections of insulin. Later with a diet change, she was able to control this problem, with oral medication.

Art was one of Ira's major passions and she continued with her art, designs and paintings and has painted many scenes of the farm, the buildings and farm life in general. She is a gifted artist and many of her works adorn the walls of their home on the farm and later in the home that she and Clem had built at Mandurah.

The bread was still being delivered by the local baker from Serpentine. Mr Maurie Mosedale would drive his red bread van and deliver bread around the district a couple of times each week. He was a fairly colourful character and was usually intoxicated with liquor on his return trip. Lorna recalls a time when she and Peter Murray, a neighbour, were riding their bikes home from school at Keysbrook when they heard the baker coming up behind them. When they looked around he was heading straight for them. They grabbed their bikes and got off the road and up the bank on the side to avoid being hit. They moved just in time as he sped past almost touching them, then he swerved back across the road and continued on his way again. A few years later he was killed in a car accident on Kerosene Lane in Baldivis.

Cliff Tomlinson is another person who used to visit the farm regularly. He was the victim of Polio in his younger days and had lost the use of his legs which had withered. He had a special wheelchair and would board the train in Perth and then be collected from the Keysbrook Railway Station. He spent many happy hours with the families. Cliff was quite a character and would sometimes encourage the girls to say grace in a different way which resulted in them getting into trouble. "Little fishes in our dishes. Hallelujah, amen" was the little pseudo prayer that got them into trouble. Another joke he would play was during the washing up after tea. He tied the corner of the tea towel to the handle of a china teacup and the throw the cup towards you so that you just couldn't catch it. He held onto the other end of the tea towel so the cup would come back to him. Just as well that we had good cups in those days.

Because his legs were of no use to him he would fold one over the other and scoot along the floor or path using his hands and bottom. He would get around in this manner very well. When he was visiting Lance and Vera and family at Mount View he used this method of mobility to climb nearly to the top of the hill opposite the house. He eventually got a motorised wheelchair and this allowed him a lot more freedom.

Lance and Vera and family spent a few holidays at Albany. At the time when Neil was a baby, they would all pile into the Ford Prefect that they had at the time and off they would go. They had a break in the journey at Kojonup where they stayed for the night at Vera's sisters, Margaret Lambie, place. Next day away they would go again and they stayed at Emu Point. Swimming and fishing were favourite pastimes and there was a small boat involved. A small dinghy, which was named "Tomtit", was moored in the rushes close to the edge of the water. Occasionally the girls would take the boat for a row in the estuary. One day while manoeuvring the dinghy to tie up at a jetty Marjory lost her balance and got her finger jammed between the boat and the edge of the jetty. She had a very painful finger for a few days and this spoiled her holiday. A few days before the trip to Albany the girls were playing bowls in the passage of the "Wee Hoose" when Lorna had a turn and got her hand too close to the wooden floor and a large splinter lodged under her fingernail. The nail came off while they were at Albany.

Clem and Lance were interested in a block of land along Hopelands Road west of Keysbrook. In 1952 they purchased this "The Block" of three square miles (1920 acres). Although the soil was of a fairly sandy type it carried a good stand of bush. In the middle of the block were several swamps with a fairly good track around them. Around these swamps were many paperbark trees, tea tree and swamp grasses. On a few occasions, the families would go for picnics to this swampy area as it was a pleasant change from the cleared land where they were grazing cattle. The girls, Gwen, Marjory and their friend Isabel Sutton would take the horses Trixi, Pharlap and Sandy down the block and race them through the water. The bush provided a great challenge to ride horses through and they had many a pleasant time here away from the main farm. A well was dug and Lance and Marjory installed a windmill and pump to provide a water supply for the stock. Bush poles were used to construct a hay shed and hay was stored there for use in the summer when paddock feed was in short supply. David had an experience with a snake when he jumped down from this haystack and landed on the ground very close to it. When he screamed out in fright his father came and took care of the snake. He was about 5 years old at the time.

Soon clearing would begin and these swamps would disappear but it does now provide some very good pasture and hay.

Lance and Vera travelled to Geraldton for a trip and holiday. While there, they visited the Wicherina dam which was covered with a roof. This was the largest covered dam in the Southern Hemisphere. Whilst at Geraldton Lance developed asthma, probably from the grain dust and this condition continued with him for the rest of his life.

In the last few weeks of school in 1954, David visited the Keysbrook School with Lorna to see what it was like being at school as he was to attend the next year but this was the year that the Keysbrook State School was to close for the last time. The building remained on site and was used for many years by the local women as their C.W.A. club rooms.

Most of the local women attended Country Women's Association (sometimes called the Chin Waggers Association) meetings each Wednesday. They would get together for discussions, needlepoint work and other handy crafts. This was a great outlet for them as most were busy on their own farms and had little other contact with those people outside their farms and families. The women often made little gifts for each other, particularly at Christmas. Enid was not very impressed when she received such a gift from Vera, Ira or Doff (Mrs Les) Robinson as she thought that they should be able to afford to buy a gift.

Marjory was attending Armadale High School for her first year in 1954 and was required to ride her pushbike to Serpentine to get onto the Armadale School bus driven by Mr. Bill Fitzgerald. She was leaving home about 7:00 in the mornings and returned home about 5:00 in the afternoon. It made a very long day and added to the difficulty of study.

Close to the edge of the Bunbury Road, just in the paddock was a giant Morton Bay fig tree. This tree-covered a large area and was an ideal place for a cubby. Lance and Vera's children spent quite some time playing under this tree. One of the games was to tie a string around a block of wood wrapped in brown paper to make it like a parcel. This was then placed in the middle of the road. When a motorist stopped to pick it up the kids would pull the string and the motorist thought that the parcel had disappeared. We used to put the address of 60 Wyatt Rd, Bayswater on the package but don't think any got delivered.

The kids were very fond of Mills and Wares ginger nut biscuits. They often had a few in their pockets. Neil lifted up one of the old bags that covered the floor of the cubby under the tree and found some worms. He liked playing with the worms so he put some in his pocket. Later when we returned home someone asked if he had a biscuit in his pocket and when he put his hand in, he pulled out a mess of biscuit and worms. From this day on he had the nickname of " Grub ".

Pop Gomm made a small box type cart fitted with a pair of lightweight shafts and bicycle wheels. This was fixed to a pet billy goat with a harness and the kids would have rides and a lot of fun. Sometimes an old pet sheep named Buster would be harnessed to the cart.

The numbers at Serpentine State School increased in 1955. Because of the closure of Keysbrook School, some pupils from there were now required to attend Serpentine. Lorna and David along with Joan and Edward Ingpen. The Government was providing a bus to collect children from outlying areas and take them to school and return them home again in the afternoon. The bus collected the children at about 8:00 am and had them home again close to 4:00 p.m. It was an old Ford bus and usually driven by Mr Ted McMillan of Serpentine.

Originally the bus route took the bus south from Serpentine along the Bunbury Highway to Page Road, then back to Keysbrook, then west along Elliott Road and right into Wescott Road. After turning left at Utley Road it then turned right into Rapids Road. When it reached Karnup Road it turned right and continued to Serpentine where it crossed the railway line and turned right into Richardson Street then left around the Pub corner into Wellard Street then right into Lefroy Street where the Serpentine State School is situated.

In the late 1950s, the bus route was altered to serve the people of the Mardella area. When the bus reached the intersection of Rapids Road and Karnup Road it continued across the intersection and then turned right onto Lowlands Road and then over the railway line, turning right into Wright Road which becomes Richardson Street when it crosses Summerfield Road, then onto the old route to the school. This extra distance coincided with a new Austin bus which travelled faster but still added about 15 minutes to the time taken. Mr Green of Serpentine was a regular bus driver of the day.

Lance became ill and was admitted to St. John of God Hospital in Subiaco for some time. Gwen was only 16 or so and had to operate the hay baler for that season. Marjory and her friend, Diane O'Leary, went to visit him. They dressed in their smartest clothes, put on lipstick and a flower in their lapels. When Lance saw them with their make up on he gave her such a dressing down as he considered her far too young to be painted up like that.

It was in 1955 that Clem was made a Justice of the Peace, with Lance being made a Justice of the Peace some ten years later. This gave both men a greater sense of Civic Duty, as part of their duties was to act as magistrate for the Local Court. Many people within the district used their position for the witnessing of signatures when signing legal documents.

Lance was concerned with the direction that the Dairy and Milk industry was heading and decided to use his experience to assist with the management of the industry by becoming a member of the Farmers Union. Soon after joining this organisation, he was elected to the Executive Committee. A position he was to hold for the next 18 years.

New Beginnings

Clem and Lance were proceeding with their plans to operate the two separate farms and Lance took the opportunity to call tenders for the construction of their new home. Mr Bob Stidwell was a young budding architect from Gosnells and was retained to draw up the plans for the brick and tile dwelling. Lance selected Plants of Waroona as the building contractors as he had seen some of their work previously and was happy with it. The house was completed for habitation in June 1955 and was fitted with all the "mod cons" such as 32-volt power and a toilet built onto the back verandah. A small shed was constructed to house a diesel engine, generator and a set of batteries.

A well had been constructed some years ago with a big windmill fitted to it. This supplied water to the house and for the stock. There was a lot of "coffee rock" in the well and this caused the water to stain everything an orange-brown colour. This made the water unsuitable for the house so a large concrete water tank was constructed to hold the rainwater as it came off the roof.

David and Neil had an experience that would have far-reaching effects on their future. Lance took them with him when he drove over to Bolinda Vale to help with the milking. He put David and Neil in the boot of the Austin A40 car. The reason for this is not clear. When Lance arrived at the dairy he had his mind on other things and left the car and went directly to do the milking, inadvertently leaving the boys in the boot of his car. The boys thought that he was having a game and would be back soon but after some time they became agitated. However hard they kicked and banged on the boot lid they could not make themselves heard and they became very upset and distressed at being locked in such a confined space. It wasn't until the milking machine was turned off after the completion of milking (about two and a half hours) that someone who was walking past the car heard the noise and told Lance. He exclaimed, " Oh my goodness the boys are in the boot, I completely forgot them". He immediately unlocked the boot and out came two very upset and frightened young boys. To this day both David and Neil have extreme difficulty in containing themselves in a small enclosed area. The experience has left them with a feeling of strong claustrophobia.

Both Lance and Clem had property in the hills and this was becoming increasingly difficult to fertilise by hand. The other option was to use aeroplanes for the purpose. These were expensive but still the most cost-effective way to do the job. At the end of Fisher Road in Fletcher's paddock was the first airstrip as the old Tiger Moths required a fair distance to get enough height to clear the hills. This strip was used for several years and at one time they used a small strip just over the creek from Mount View homestead across the Bunbury Road from Karnup Farm. When we returned home from school on the bus one day we saw this aeroplane tail poking up in the air with the nose of the plane resting on the ground. Apparently, a gust of wind had caught the plane as it was preparing to load up. It was blown across the strip and through a fence. During the 1960s, Clem and Lance along with some other locals constructed an airstrip up in the hills near Scarp Road. This made the operation much faster as the planes were already on top of the hill when they loaded up. Clover seed was added to the super prior to spreading and now there is a good body of clover through the hills blocks, providing good stock feed.

Lance and Vera had both being providing Sunday School lessons at Serpentine for many years and Lance was made Superintendent (His father had held this position in South

Australia). While Mr Laurie Manning took the Sunday School at Serpentine, Lance and Vera would travel to Mundijong and hold Sunday School in their local Church. Lance was also quite busy with Lay Preaching and would frequent Jarrahdale, Pinjarra, Waroona, Mandurah and Byford churches.

Neil began his education in 1957 when he attended Serpentine State School with Lorna and David. The following year Lorna would commence her secondary education when she attended Armadale High where Marjory had just completed her secondary education.

Marjory had finished high school in 1956 and started work on Bolinda Vale in January 1957. She and Gwen worked together with their father and uncle until Gwen left just prior to her wedding. One of Marjory's favourite jobs was to get the cows in for milking. She would saddle up the horse, usually Sandy and sometimes Trixi, and ride around behind the cows and drive them to the dairy. On one wet day, she decided to keep herself dry and wore a raincoat. Unfortunately, it flapped in the breeze and this startled the horse, Sandy, and he bolted across the paddock with Marjory clinging onto the back of him. She tried to fold the coat tails under her bum to sit on them but every time the horse moved, she lifted off the saddle and the coattails came loose again. This frightened the horse a bit more and off he would go faster again and the faster he went the more noise the coat tails made. After 15 Ä 20 minutes she was able to get him under control and after removing the raincoat walked him back to the dairy. A hairy experience to say the least.

Lance and Vera's eldest daughter Gwen was married to Mr Fred Swaby on the 1st of June 1957. Fred had a mixed farming property to the west of Byford and they went to live on this farm. They bought and sold and worked several farming properties until they purchased the Kelmscott Caravan Park. They progressed from here to the Perth Tourist Caravan Park. Their children were a big help to them with the running of these caravan parks and it was a real "family affair". When they sold this they went onto caravan manufacturing and transporting vans to and from the Eastern States. They have now retired and after living in a house on the canals in Mandurah have returned to live at Byford.

In 1957 Clem stood for and was re-elected to the Serpentine Jarrahdale Road Board. He kept this seat until 1961.

Ira and Clem took the opportunity of a holiday in 1958 when they boarded the "Southern Cross" bound for South Africa. They were away for about 4 weeks and visited Durban and experienced the natives in their "whirley" huts. Capetown was the terminus of the holiday and they travelled from here to Johannesburg and Pretoria by the Blue Train. Clem viewed the settlements set aside for the natives and is of the opinion that although there was Apartheid in South Africa they were better looked after than the Aborigines were in Australia.

While at Johannesburg, Clem and Ira were escorted over one of the experimental farms where they were working with different breeds of cattle and also with Artificial Breeding. Kruger Park was a highlight of their trip as with their early morning start they were able to see nearly every type of wild animal before they headed for shade for the hottest part of the day. The return trip was on the "Northern Star".

A friend of Clem and Ira, a Mrs Knobs looked after the two children while Clem and Ira were away on their trip.

In 1959 Marjory was to begin her Nursing career. She had worked fairly hard on the farm and was happy for the change. Marjory studied at Fremantle Hospital and was home frequently on the weekends and for some of her holidays. She studied and worked at several hospitals around the state. In 1963, on September the 21st, she became married to Greg O'Neill and spent some time at Lake Grace, Newdegate and Beverley. She, Greg and their four sons spent a few years in East Africa and later returned to the farm. She was unfortunate with her marriage and remarried after her divorce. Ken Peters had a fuel delivery business in Forrestfield and later a farm at Brookton, then at Pinjarra. Marjory continued with some part-time nursing to supplement the farm income but the partnership did not work. She remarried again and is now living happily with Ron Tree at Quairading.

Lance and Vera took a driving holiday to Tasmania in January 1959. After driving their Austin A90 to Melbourne, they put the car on the boat and departed for Tasmania. They spent 2-3 weeks there and had a good rest and looked at some dairies as well. Mr Joslyn was one of the people whom they stayed with while they were there. As they left W.A. they were experiencing a heatwave and this followed them for all the time that they were in Tasmania. Upon their return, they began designing a new dairy.

Clem and Ira's daughter, Coralie began her schooling in 1959 when she attended Serpentine State School. She was able to travel on the bus that was provided and was picked up from and returned to, the front gate of the farm.

Vera always had lovely long black hair that was always plaited and tied around her head. She and David were trying to get a temperamental lawn mower to go. David was pulling on the rope starter and his mother tried to help him. Unfortunately, probably due to lack of experience, she gripped the belt and when it rotated around the pulley it pulled her fingers in as well and severely damaged two fingers on her hand. She was taken to hospital for attention but the fingers were bad for a few months. Because of the discomfort of the damaged fingers, she was not able to do her own hair and after having someone else do it for her for some time, decided to get it cut off to a style that she could manage.

During 1959 Lance and Vera began the construction of the new dairy in preparation for the forthcoming dissolving of Kentish Bros. and subsequent splitting up of the dairy herd. They selected the most modern design as they had visited various dairies during their recent trip to Tasmania. The design allowed for the cows to step up onto the milking level while the operators worked at the lower floor level. This made milking a little easier on the back.

The dairy was constructed by Green and Saw Bros. from Armadale and the welding fabrication was completed by Barry Jubb of Jarrahdale. A lot of care was taken to make sure that the floor slopes were correct and the drainage was adequate. With a further elevation of the milk room, the job of lifting full cans onto the truck was much easier. A diesel engine was used to provide power for the vacuum pump and also for the 32-volt generator.

Lorna completed her education at Armadale High in November 1959 and immediately began work on the farm with Lance and Clem. She fitted in well with the dairy and stock work.

Lance was becoming increasingly concerned with the less fortunate youth around Perth. He was on the Board of Directors for the Mofflyn Children's Home in Victoria Park with the Methodist Church. They set up a farm at Werribee (near Wundowie) where boys would be given housing and care whilst they were instructed in farming activities and practices. This also involved some farm work which was very beneficial to them as they would be better absorbed into the community. However, there are some people around who did not agree that these boys should be made to work, albeit for their own benefit. When the boys were no longer permitted to work, others needed to be employed thus causing the operating costs to increase. The Werribee operation had to be closed down and the farm sold.

The dairy herd continued to grow and in the year just prior to the herd being divided, they were milking 250 cows twice each day. At the time, this was the largest dairy herd in Western Australia and they were producing about 2000 litres per day to fill their whole milk quota contract.

The days that both Clem and Lance were dreading had arrived. This was the time to divide not only the land but also the dairy cows, machinery, tools and other stock. For years they had worked together amicably and now with the dividing up of everything, that period of their lives had come to an end. It was a very stressful time for both men to divide their belongings and go their separate ways but they had planned for this day and knew that it was the correct thing to do. There had been no animosity amongst either of the men or their wives although sometimes things did tend to get a little difficult. Their Christian upbringing had stood by them and they overcome those little problems in the most Christian manner.

Now that they were working separately each would need to learn to be more proficient in the activity that the other had been better at. Lance learnt a lot more about stock work fairly quickly as did Clem learn more about the mechanical aspects of farming.

Each of them took about 125 dairy cows each along with a quota of about 1000 litres per day. The cows of Lance and Vera's were driven along the road after the mornings milking to their new paddocks. Now the fun was to start. Lorna remembers working on Bolinda Vale after she completed her final year at Armadale High School. She helped sort out the cows and drive them along the road to Mount View. Marjory was home for a weekend from nursing at that time and she helped sort out the problems that they had teaching the cows to step up onto the raised floor in the shed. Most of the older cows performed remarkably well but some of the younger ones didn't like the change and balked at the step. With some careful handling, they overcome the problem but production did suffer for some days.

During this same year, Lance was elected to the position of President of the Whole Milk Section of the Farmers Union of Western Australia. He was prompted to this position by the crazy widespread prosecutions of dairy farmers on the debatable question of milk quality. Lance would attend meetings in Perth and regionally very frequently and held this position for 8 years.

The Henderson family were working on Mount View in the 1960s and Mrs Henderson was a very large lady. Lorna had a pet budgie in a cage on the back verandah and had just taught the bird to do the wolf whistle. Mrs Henderson needed to speak to Vera and had walked up from the house near the railway line. Just as Mrs Henderson came through the side gate she was greeted with this perfect wolf whistle. She was not impressed.

She was quite flustered from the walk up the paddock and to be greeted like that was just not on.

Colin began his schooling in 1961. He attended Serpentine Primary School along with his sister Coralie along with his cousins David and Neil.

The old house was showing signs of its age when Clem and Ira began building a new brick and tile house close by. They designed the house with 3 bedrooms with an office and a double garage under the one roof. This new home was a vast improvement on the old home which was then used as a workers cottage and subsequently converted to the Museum.

In February of 1961, there was a major fire in the hills area. For a week firefighters battled against the advances of the fire beast but it eventually covered an area from Kingsbury Drive to Pinjarra and east for many miles. Things were looking fairly grim in the district and Marjory was called back to the farm from her nursing to assist with the dairy while the men were attending to the fire. Dwellingup township was razed completely and several people lost their lives. Lance, Clem and many others spent 15 or 20 hours per day in the attempt to help bring it under control. One night, the fire was approaching the Bunbury Road from the east, so a back burn was started along the highway in front of Mount View. As this was burning and encroaching up the hill the east wind began to blow, causing some burning cow pats to roll along the ground and sparks from them were lighting up fires in the paddocks immediately to the west of the road. A team of men spent all night mopping up these little spot fires and saved any damage to Mount View. Lorna didn't have her driver's license at that time and the local traffic inspector, Mr Bill Fitzgerald, turned a blind eye when he saw her driving the little Austin truck and water tank.

After it was all over the Army arrived and lit their own fire to cook their breakfast. This didn't go down very well with the locals. Marjory was driving the tractor with her father standing on the water cart behind. They were putting out the small fires around the wooden fence posts and on the edge of the back-burn. The Army guys began laughing as they had obviously not seen a young sheila driving a tractor before. Gwen came to help as well but had to stay in the house as she was pregnant with Doreen at the time. Fred was out with the other men with his fire unit. So it was all hands on deck and a family affair when there was a crisis.

Lance organised a hay supply for those people who needed some and a large stack was built just east of the dairy. The unfortunate part of this is that some people who were not fire-affected also took advantage of the free hay, much to the disgust of Lance.

When Clem would take his truck to the "Block", he put his horse in the back and took Coralie with him in the front. She always enjoyed being with her Dad on the farm and always wanted to join him on the farm after her schooling. When Clem was riding around checking the stock he would frequently look at the truck to see the small head of Coralie above the dash. She would stay in the cab of the truck while he was away but was always happy to have him return.

When Coralie was about 3 years old she followed her Dad into the hills when he was doing some burning off. He didn't know she was there. Although she had lost sight of him she continued walking and eventually lost her way and got caught in the corner of the fence. She couldn't go any further so she just sat down on the ground. After they had finished their job, Clem and the men returned down the hill and just happened to stumble across this little girl who had become lost.

In 1961 the New Local Government Act was gazetted which changed Road Boards to Shire Councils and Road Board Members, to Councillors. The Chairman was now the President. All Road Board members were to retire and an election was held to elect seven new Councillors. Clem was elected, topping the poll.

In the early 1960s Clem purchased a new Grasslands Forage Harvester. This would cut the green fodder to make silage. They had tried silage before but now there was machinery to do the job and this would make the operation more economical. A large pit was constructed and the grass was dumped into this and then rolled with a tractor to remove the air. This method proved to be quite satisfactory and the operation continued for years. Lance could see the value and soon purchased the equipment for the same job on Mount View. Some years they would each make about 200-300 tonnes of silage.

Hay was still a very important feed supplement for the dairy cows, particularly to give them dry feed in the winter and to supplement the paddock feed in the summer. The hay was baled now and a mechanical loader was used to lift the bales from the ground to the truck. An elevator lifted the bales from the truck to the stack. All of the rest of the work still had to be done by hand and with the number of bales required, this was still a labour-intensive, dusty and heavy job. It was not uncommon to produce about 40-50,000 bales of oaten or meadow hay each year at each farm.

Lance and Vera bought a small bond-wood boat, fitted with an 18 horsepower outboard motor. They and their children spent many happy hours at Rockingham and Palm Beach playing with the boat and learning to water ski. This was later updated to a larger boat with a 40 horsepower outboard motor and now they could ski much faster and continued to have fun as a family. This use of the boat was usually done between milkings on the weekend and during the two weeks annual holiday at the beach.

Lance's cousin, Keith Nicholls, his wife Kitt and their two sons Don and Peter were very keen skiers and boat people and spent many hours together enjoying the sport.

Lance and Vera increased their holdings on September 1961 when they paid a deposit and agreed to purchase a 74-acre property from Mr Atkins. This block adjoined "Hectors" and "Fishers" and was referred to as the "New Farm".

David attended Pinjarra High School in 1962 and 1963 for his secondary education. By riding his pushbike to the Keysbrook store where he caught the bus to Pinjarra. Starting out at 7.30 in the morning and returning by the same route in the afternoon, arriving home at about 4.30. Edward Ingpen was at that school at the time

In the early 1960s the Government was involved in a program to provide electric power to the south-west of the State. Mr Arthur Farnham was the electrician who carried out much of the initial installation work in the district and was kept very busy. Most farmers would erect their own poles and he would come along and fit the cross arms and insulators onto them and carry out the necessary wiring and connections.

Lance and Vera had the SEC power connected to their house on June 1st 1962, with Clem and Ira having theirs connected two days later. Until this time they were relying on a 32 Volt DC system. Now both Vera and Ira needed to purchase those electrical appliances that were necessary for the kitchen that we take for granted now. These included a washing machine, Mixmaster and electric iron. She would add more appliances to this list as time went by.

Power was connected to the dairy on Mount View in June 1962. This made working in the dairy a little easier as the noise of the diesel engine was replaced with the quietness of an electric motor. There was still the noise of the vacuum pump and the pulsators but this did make the job more bearable. The lights were much brighter as the old 32 Volt lights were fairly dull. Because of frequent breakdowns with the new power supply, the old engines were left in place to operate the dairies when the power went off.

Lance was honoured by the Judiciary of Western Australia when in 1963 he was bestowed with the honour of Justice Of The Peace. Among his other responsibilities in this position, he also was deeply involved with the establishment of the Karnet Prison Farm along with Clem. They sat on the Bench of that Prison for many years.

November of 1961 saw Lorna begin her Mothercraft Nursing at N'Gala. She studied nursing and worked at N'Gala until March 12, 1963. While she was at N'Gala she became engaged to a young chap from Armadale who was working as a cabinet maker and running his small orchard.

Lorna married Barry Robins on the 26th of October 1963. After their honeymoon, they settled into their new house on Eighth Road in Armadale. Barry had an orchard in the paddock to the west and north of the house and was also working as a cabinet maker. They and their two daughters moved into their new house on the hill just east of Armadale in 1984. They had sold their house in Eighth Road and subdivided the orchard, sold some as housing blocks and built houses and duplexes on some blocks for renting. In the early 1990's they built another new house this time at Mandurah to where they intend retiring.

Lance and Vera purchased for themselves a caravan in November of 1963. They would use it for family holidays at the beach and also for the travelling holidays that they were to become so well known for. This was the start of their caravanning and they were to continue travelling with a caravan for many years, saw most of Australia and met many friends on the road. They instilled the travel bug in their children as they also followed with the practice of holidays and trips with their own caravans.

Clem was elected President of the Serpentine Shire Council in May of 1964 and he held this position until 1989. He was very busy with this position but he was careful to balance his time so that neither his farm nor his office suffered from lack of attention.

While Neil spent his first of three years of high school at Pinjarra in 1964, David was sent to complete a 2-year course at Narrogin School of Agriculture. Lance wanted his sons to get an insight into different types of farming and to investigate some new methods of farming practices.

Heather turned six years of age in 1965 and it was her turn to begin school. She also attended Serpentine Primary School and travelled on the bus. Coralie began at Pinjarra High School in 1966 where she was to stay until completing her 3 years of secondary education.

David completed his education at Narrogin Agricultural College and returned to the farm late in 1965. He was very articulate with his hands and proficient at mechanics and welding so he had plenty to do around the farm. Most of the construction and mechanical work was now done by him and this allowed his Dad to get onto to other projects. Staff were employed in the dairy and with either Lance or David working with them the other had the chance to spend more time doing other farm work. Many of the jobs which had been left undone could now be done and things progressed well. Lance could see the opportunity to expand and kept his eye out for any suitable land for sale.

In 1966 a property west of Mount View down along Utley Road came on the market. This was owned by a Mr Arndt who operated a small dairy on the property. Utley road divided the farm into about 200 acres on the north side and 100 acres on the south side. Most of the farm was a mess with wire everywhere and broken down fences and buildings. The property was purchased and we went to work to clear it up. Neil was in charge of the bulldozing contractor to tidy up an area on the north boundary. Mr. Storey was employed to do most of the fence repairs. Neil and David spent a week or so and put down a well by hand on the north side to provide water for the stock. This was heavy back-breaking work and it gave them a good background knowledge of how their Dad and Uncle had worked in their younger days. Concrete well liners were used and the well was equipped with a jack pump and electric motor.

Lance was beginning to slow down and David was able to carry a greater load of the work. Some new technology which he had learnt at Narrogin was put into practice, particularly with the more efficient distribution of run-off water. At the block at North Dandalup, there was a drain letting water into the paddock which came from the road. Over the years this had cut quite a channel in the ground and made part of the paddock unusable. By banking off the drain and constructing some carefully measured grade banks across the paddock they were able to stop any further damage to the channel through the paddock and also rejuvenate a sandy hill. When the dam was enlarged some of the spoil was carted into the channel and now the whole paddock is useful again. This same practice was adopted on Mount View and Neil has continued with the practice and it has been very beneficial to the production of the farm.

David's asthma was affected by the wet winters and also during the spring. Some days he was too ill to work and some days he wished he didn't but he survived. Mowing for meadow hay was particularly difficult as he got to the stage where his eyes would close up and he could barely see. The doctor prescribed drugs to offset the effects but although these made him sleepy he needed to take them. The oats dust in the dairy was very bad and on some days and he wore a breathing apparatus. After a few months he learned to live with the problem but there was still a lot of discomfort.

He had many discussions with his father about the methods he was using as David had been shown more up to date methods at Narrogin. This caused some conflicts as Lance was fairly set in his ways and was difficult to convince of change.

Thursday evenings after tea was set aside by David to go to Youth Club in Serpentine. He attended here for several years and made some close friendships. They occasionally would organise Saturday night dances and car rallies to give themselves something to do outside of farming.

Clem and Ira took their three children off to see New Zealand for a three week holiday in the mid-1960's. They flew over by plane to Auckland and hired a car there to get around both Islands. Clem was very impressed with the Angus cattle but Ira said "We see enough cattle at home. Let's have a holiday". So they did. They saw their first snow and glaciers. They were all completely taken with the flow of water in the rivers that was created from the snow and ice melting in the high country.

Lance and Vera were becoming really modern when they purchased their first television set in November of 1966. Previous to this they relied on the newspaper and the radio for the news of the day and some of their entertainment.

Towards the end of 1966, Neil completed his education and he also came to work on the farm with his father and brother. The boys and their father worked together with Neil being able to fit into the jobs that David wasn't able to do.

The two boys worked with a group of others to construct a go-cart track in Serpentine where the old timber mill had stood. Between them, they bought a competition Go-Kart and had a lot of fun.

The cost and unreliability of labour was of great concern to all in the industry. Clem decided to tackle the problem head-on and build a new, modern dairy in the attempt to become more efficient. He selected the most advanced design for the times and build a substantial dairy using the "Herring Bone" design. This allowed them to operate the dairy with less staff, so reducing some labour problems and allowing more efficiency to the operation.

David became engaged to a young lady in May of 1968. They met whilst attending and helping at the Serpentine Youth Club. Barbara McKay was the Manageress of "Tom The Cheap" grocer in Pinjarra. She is a very capable person and a hard worker who finds most jobs not difficult to accomplish. They enjoyed similar interests and activities and worked well with the go-cart club, Serpentine Youth Club, and the Methodist Youth Group of the Church in Mundijong. Barbara's parents had a beach cottage at Mandurah and David would visit there on many weekends. They became married on October 19, 1968, in the Methodist Church at Byford and held their Reception in the Town Hall at Armadale. Barbara was attended by Judy Nairn (Byford) and Elaine Redfern (Mandurah) as her bridesmaids. David had Robert Hardey as his best man with Robert Manning as groomsman. They returned to the farm for the hay season after a 3-day honeymoon at Mandurah.

Coralie began working on her father's farm when she completed her secondary education at Pinjarra High School at the end of 1968. Her first jobs included the washing down of the dairy after the milking had been completed and the feeding of the calves. She had a cart made to carry a few milk cans. One of the men would load the cans onto the cart and she would take it to the calf yards. She dished out the milk for each calf individually until they had all been fed. Usually, up to 60 calves would be fed twice a day. At 4:30 in the mornings, Coralie started her day when she would walk up into the hill paddock to bring in the dairy cows for their morning milking. After the milking was done and when the dairy washing was completed she would walk the cows across the Bunbury Highway and the railway line for the days grazing in the paddock. There were always several employees to attend to the main work but these seemed to come and go and consequently more and more work was carried out by Coralie.

Modern methods of farming were being practised by many farmers and one of the latest practices to take hold was the bulk handing of fertiliser. Lance bought a super bin and a new Dodge truck in December of 1969 to cart the superphosphate for the farm. Previous to this all the fertiliser was handled in bags and this was very heavy work as the bags weighed about 84 Kgs each when full and a fair bit of handling was necessary.

Clem's farming experiences have enabled him to have a sixth sense when it comes to daily tasks. He is always aware of what needs to be done even well before it is necessary. Coralie is developing this sixth sense but she says that her Dad still leaves her in the dark. This is a very important attribute to have when operating a farm of this nature.

The turnover of staff was still a constant problem for all in the dairy industry and Clem was not exempted from this problem. Many dairy and farm workers were of Dutch descent and some of English descent, some good ones and some bad ones. One of these workers of English descent appeared more stable than most and stayed for quite a few years. He became quite friendly with Coralie and on the 7th of July in 1973 Graham Parkin and Coralie were married. Rosemary Della Franka and Heather attended Coralie while Neil was Grahams' best man. Graham's cousin travelled from England to be his Groomsman. After the ceremony in the Pinjarra Methodist Church, they travelled to the Shire Hall in Mundijong for their reception. They boarded an aircraft after the reception and spent their honeymoon in the United Kingdom. Here they stayed for some time around Bingley in Yorkshire, where Graham's people lived.

A house was constructed on the property across the railway line from Bolinda Vale, close to the creek on Dirk Brook Road. Coralie and Graham have since added sheds and a garage to the vicinity of the house and are now using this as the base for their farming operations.

Clem, Ira, Coralie and Graham formed a partnership and continued with the efficient operation of the farm. They could foresee that they would benefit greatly from beef cattle so they purchased more good breeding stock and worked on building up a quality beef herd.

David and Barbara with their son Kevin left the farm at Mount View in September 1970 and began their lives away from the farm. Their intention was to find a more suited climate for David's health and settle down. After spending some time at Mandurah they headed off to Carnarvon where David worked in the building industry and then got employment with a drilling contractor who was working in the area but was based in Geraldton. At the completion of the contract with the Government, they moved to Geraldton. David and Barbara purchased part of Peter Beaton's plant and began their own business with boreholes, windmills and pumps. They developed a large business there and sold this in 1988. They moved back to Perth, where David and Barbara worked as employees for a time before buying a property in Williams. After moving to Williams they began constructing a piggery, which they operated for 4 years. During this time they were producing up to 55 prime porkers each week and selling directly to Woolworths. They then leased out the farm and piggery and moved back to the Perth area again, to their new home in Forrestfield.

Lance and Clem had been shifting cattle across the Bunbury Road and the Railway line for many years. There have been the occasional problems with some road users causing a few close shaves with stock nearly being hit. The railways ran to a tight timetable and by being careful, problems could be averted. However, in July 1970 a train was running out of schedule and collided with a mob of cattle being driven across the line. The impact killed 11 of some of Clem's best dairy cows.

Clem and Ira's son, Colin was a good scholar and very capable with his hands. He was frequently working in Clem's workshop mending and making things. He used many of the hand tools and electrical tools to carry out the projects that he worked on from time to time. On Sunday 20th September 1970 he was working in the workshop on one of his projects when he had an accident. The grinder wheel disintegrated and he received a very bad blow to the forehead. It was some time before he was found lying on the ground but was then rushed immediately to Royal Perth Hospital for attention. After receiving emergency attention he was admitted for further treatment and observation. He did show some signs of improvement but wasn't shifted to Shenton Park Annex until 30th October.

The injury that resulted from this accident has left Colin with a communication deficiency and some epileptic problems. He underwent several operations to remove scar tissue from the brain which had caused pressure to build up thus causing epileptic problems and a great dependency upon medication. Whilst relieving the problem, the surgeons were not able to solve it completely and Colin continued with a reduced medication regime to help him control his communication and epileptic problems.

The handling of hay received a boost in the 1970s when a motorised hay wagon was put into use. Now one man could drive this unit, pick up a load of hay and stack it into a large stack without having to handle any bales by hand. The feeding out of hay was still very manual requiring quite some time and effort.

Even with the need for staff being reduced, Clem was still having problems with them and this was a big factor when they made the decision to quit dairying during 1974. The whole milk quota was sold along with the dairy cows and then they could concentrate their time to the herd of beef cattle that they had been building up for quite a few years.

In 1974 after she had completed her Secondary Education at Pinjarra High School, Heather gained employment in Perth. She wasn't the farming type and was determined to make her own way in life. She became married to Colin Payne, a car salesman, in September of 1986. After living in the City they came back to Bolinda Vale to live in the house ("The Wee Hoose") which was originally built for Lance and Vera.

This house showed some signs of deterioration. Graham replaced some of the stumps and they did some repair work to the rest of the house and gave it a coat of paint, prior to Heather moving in.

Lance, Vera and Neil were operating the dairy with the help of employees and were not happy with the direction that they were taking, so they made a decision to quit the dairying operation. They were able to sell off their whole milk quota and had an auction sale for their dairy cows in Mundijong. They had already begun building up a herd of beef cattle and with the purchase of some sheep, they continued farming but in a different manner. The houses that were used for staff housing were now free, so they were able to rent these out which added to their farm income.

The Serpentine Jarrahdale Shire built a new Hall in Serpentine to replace the old weatherboard structure, this was completed in 1979 and named the "Clem Kentish Recreation Centre ". This was officially opened by Clem as the Shire President. In that same year, he was also presented with a Gold Medallion for having given over 30 years of service to Local Government in Western Australia.

Deborah began her education in February of 1980 when she began attending classes at the Serpentine Primary School. She would walk to the highway and be picked up there by the school bus that took her to and brought her back from school.

During the early 1980s Clem and Ira took a trip to holiday in Europe. It was during this trip that they met up with Kathy and Helen, the two American ladies who became very good friends. They spent some time in Yorkshire, visited Loch Ness, called in on Joy and Trevor Moffat (Esther's daughter and family) in Scotland. They left the car in London, flew over and toured the Scandinavian countries, then returned. They caught the tour bus to Dover and crossed the channel in a ferry and continued on by coach to Belgium, France and most of the other European Countries. The tour took them on to Italy and Rome, the dirtiest city they had seen. Paris was another city that Clem wasn't impressed with as the natives were not very friendly.

Clem was having quite some trouble with the deterioration of the joint in his hips. This was causing him a lot of discomfort and he approached the doctor, his brother in law Dr Lynn Smith, to investigate having the joint replaced with an artificial one. After a period of dieting to lose some weight, the operation proceeded with great success. The specialist who carried out the operation, Dr Webb, told Clem that during the operation he noticed that there was very little fatty tissue present, it was all muscle. He was to have the same operation on his right hip five years later.

With Neil and Jacky's wedding approaching, Lance and Vera began building a new home for themselves in Armadale. This was constructed of double brick and tile. Barry and Lorna provided them with a lot of help in the design and construction of the house.

On November 23rd 1980, Neil and Jacky were married in the garden of her family's home in Forrestfield, with the reception being held at the Gypsy Baron Restaurant in Armadale. Jacky was attended by Helen (from England) who was her Matron of Honour. Neil had the assistance of Ken Elliot as his best man. After their honeymoon, they settled into the home on Mount View that Neil had lived in with his parents, brother and sisters.

Although Lance was living in Armadale he still went to the farm most days to help Neil with what jobs were necessary. As the years progressed his ability to work reduced and Neil was able to do more of the work by himself and with employees from time to time. One of Lance's favourite jobs was to use his ride-on mower to clear the grass away from fences and buildings. He had an accident and cut his fingers when he tried to remove some sticks from under the mower. The fingers never mended properly and were a problem for the rest of his life.

Neil had been given some of Lance and Vera's holdings and he continued to work the farm and was leasing the balance of the properties from his parents. He bought himself a bulldozer and a road grader and with some careful measuring continued to improve the production of the land by constructing contour and grade banks to more effectively distribute the water around the property. This has proved quite satisfactory as now he has less wet areas and most of the dryer parts of the paddocks now have a higher moisture content and produce a better quality pasture.

In 1982 Clem was elected to the position of Chairman of the Metropolitan Ward of the Country Shire Councils Association for a 2 year period. He represented the Metropolitan Ward on the Executive of the Country Shire Councils Association. He had also served as a member of the Bushfires Board for quite a few years. At the same time, he was a member of the Serpentine, North Dandalup and Murray Rivers Advisory Committee for many years. A pretty busy bloke.

Neil is working with his beef cattle but has some sheep as well and needed a shearing shed, so he converted the dairy for this purpose. The raised concrete floor was removed and pens, using a grated floor, were fitted over a section for the sheep. The rest was covered in plywood flooring to form the area where the shearers would work and for the wool handling area.

Some of the sheep which Neil purchased developed footrot and he was unfortunate to have his farm under quarantine for some time. With a lot of cost, work and careful management he was able to eradicate the problem.

The sheep were required to have their feet trimmed and then bathed in a special preparation to combat the virus. This operation was difficult in the winter when the feet were soft but in the summer, hard sheep feet are very difficult to trim. A special foot bath was constructed and filled with a solution of zinc sulphate, then the sheep were walked into it and allowed to stand for an hour. This killed the virus that causes the problem. This treatment regime cost $ 3 - 4 per head and with low market prices, it was a financial burden that you could really do without.

A good return has been indicated in the future for the timber that could be produced by "Agro-Forestry". This is the new term used for growing trees on agricultural land that will be harvested for a cash crop in the future. Neil proceeded to plant some areas of the farm to eucalyptus trees with the anticipation of doing some harvesting in about 10 years. Exceptionally dry summers and grasshoppers were a big problem and a few trees did not survive but the large majority did. In addition to trees for cash crops, he has also used tree plantings to provide shelter belts within the farm. He commenced this program in 1983 when he planted about 6,000 trees of various varieties of eucalyptus, including ~ Globulus, Grandis, Saligna, Maculata and Patens.

During 1983, in recognition of his service to the community, Clem was honoured by being bestowed as the first Honorary Freeman of the Shire of Serpentine Jarrahdale.

Young Michael began school at Serpentine Primary this year. As Deborah had been attending for several years she was able to foster him through those tough times that you have when attending school, for the first time.

In the mid-eighties, further improvements were being made to the handling of hay. Now instead of making a lot of small rectangular bales, a new machine was purchased that made less, large round bales. These could be picked up with a set of forks on a tractor and now there was no need to handle the hay by hand at all. This allows a great reduction to the time taken to feed out hay to cattle and allows the farmers more time to get on with the other jobs and reduce the need for staff. Both Clem and Neil are still making some small bales for specialised feeding jobs but these are fairly low in numbers.

The two ladies who Clem and Ira made friends with on their European tour asked them to come to America. One of them had just bought a new car and wanted someone to help her run it in. Clem and Ira obliged and they had a wonderful time touring the country. They also took a trip on the train from Montreal to Banff and return. Crossed back into the States, met the President of the United States, President Ronald Reagan. Flew back to Los Angeles and took a coach tour around the State of California. They spent some time in Phoenix and on to the Grand Canyon which was very spectacular. They enjoyed the trip although they did put on a bit of weight with all the welcome and farewell parties that they attended. Kathy became ill and passed away a few years later and Helen teamed up with another friend, Lil. They came to Western Australia in 1994 for a visit which Clem enjoyed so much.

1987 saw Deborah riding her bike to the Keysbrook store where she caught the bus to attend the Pinjarra High School. She was to attend this school until she completed her education in 1991 when she returned to the farm to work in the team with her grandfather, mother and father. From this time there would be three generations of the family working together.

Neil continued with the planting of trees in 1987 when he planted out 3,000 mixed varieties of eucalyptus trees for a cash crop to be harvested from 1995 onwards. There is a fair amount of work involved in the planting of trees. First, the weeds need to be controlled so that the new plants have the least amount of competition for the available nutrients. As most of the soils have had some form of compaction over the years, it was necessary to have the ground ripped by a large bulldozer. A mounding plough follows this to form a rainwater catchment mound to assist with the watering of the new trees. Then along comes Neil with a tube-like shovel, puts it into the soil over the ripped area, drops a fertiliser pill in the tube, followed by the tree. As he withdraws the tube, he presses his feet alongside it to remove the air from the root area. Then paces out 4-5 metres along the ripped line and repeats the process. This continues until all of the area has been planted. Stock must be kept away from the new trees to allow them to grow unhindered. Usually, when the trees are about two years old, sheep can be grazed around them so from this time on, he has the use of the land again.

Lance and Vera celebrated their Golden Wedding anniversary with their family and many friends at the C.W.A. rooms in Armadale in March 1987. Lance was not well that day which was particularly unfortunate as he wasn't able to fully enjoy the fellowship with the family and friends who had turned up. Clem was up to his usual tricks and presented them with a can of Golden Syrup as a celebratory gift. Lances brother and sisters were present and most of Vera's sisters were able to attend and celebrate this occasion with them.

Josephine attended pre-primary education at the Serpentine Primary School early in 1988. This is the commencement of her formal education and she will notice many changes during the time she attends school.

In April 1988 a good contingent of Western Australian Kentish's attended the Family's 150th Reunion in Adelaide. Lance and Vera travelled over by train and stayed with Win and Hazel. Clem and Ira were on a caravan holiday in the eastern states, David and Barbara and their daughter Vicki travelled with their caravan from Geraldton. Gwen and Fred travelled over in their car and caravan. Esther and Allan were travelling around South Australia in their van. Everyone had a great time catching up with family members who have been separated by distance.

This event and experience was probably the catalyst for this book.

Neil and Jacky are members of the Serpentine Tennis Club and spend a lot of their leisure time playing and enjoying the game. They constructed themselves a concrete tennis court complete with high chain mesh fence just to the north of the house. They spent many happy hours with friends playing tennis at home.

More trees were planted by Neil again in 1988 when he planted 20,000 on the Mount View block. These were of mixed eucalyptus varieties again like before but also including ~ Uiminalis, Brookorana, Bicostata, Muellerana, Diversicolor, Camaldulensis and Calophylla. With this planting, he had the use of a tree planting outfit to assist in the planting operation. The ground was prepared as described before but now the "planter " was fitted to the rear of a tractor. The planter operator sits in the planter and drops the seedlings into the ground at regular measured intervals. The planter does the rest. This speeds up the operation and allows many more trees to be planted over a shorter time, not to mention the saving on the back muscles. These trees should be ready for harvest in about the year 2000. Lance told Neil that with the current agricultural economic climate, he would make a greater income from trees and the like in the future. It would appear that his estimation will be correct.

In 1989 Clem resigned his position on the Council of the Serpentine Jarrahdale Shire. A record of 47 years as both Roads Board Member and Shire Councillor, 1 year as Chairman of the Local Roads Board and 26 years as the Shire President of the Serpentine Jarrahdale Shire. At the age of 75, Clem needed to slow down a bit and allow some other members of the community to carry on with that work which he had been doing within the district.

1989 is the year when Stuart began his education, beginning just 25 years after his father left the same school.

Since Lance and Vera's move to Armadale, they spent many happy hours touring around the country with their caravan. Lance spent some time in the hospital with a bowel operation and later with a Prostate operation and was generally slowing down. In the late 1980's they sold their Armadale home and moved to a unit in the Amaroo Retirement Village which they purchased.

New Year's Day has been celebrated at Clem and Ira's beach shack on the beach at Mandurah for many years. Each of the family members tries very hard to attend on the day, as do many friends and neighbours. It's one of the few times each year that many of them are able to meet and this gives everyone a lot of enjoyment.

The downturn in prices for agriculturally produced products has caused many farmers to think twice about their occupation and use of the land. Neil has taken the step to diversify his production. In the past, he had seen dairying, beef cattle, sheep and wool. All of these commodities have a large fluctuation in return over the years and on many occasions, they are marginal at best. In addition to his ingress into agroforestry, he had got into an activity of collecting foliage from some of the eucalyptus trees and shrubs. These he dries and some are coated with various stains, colours and preparations to preserve them. There is a growing market for this foliage, both locally and overseas. To produce the foliage that is necessary for this type of production, Neil had undertaken a planting of about 100,000 trees and shrubs. He estimates that the 40 or so varieties will reach maximum production in about 20 years. Whatever is done in agriculture, you must always take the long view to determine the best path to take.

Michael attended Pinjarra High School in 1990 where he was to complete four years of secondary education. In 1994 he attended the Harvey Agricultural College where he completed two years of tuition. Late in 1995, he will return to the farm where he intends to work with his family.

Lance relinquished his Motor Vehicle Driver's license due to slow reflexes and failing sight. Vera would then drive her car with the caravan behind. They did a few trips in this fashion but Vera was not happy with driving and towing the caravan. Lance was becoming very slow and sometimes needed help to get around and Vera was showing the strain of the extra work. During 1990 they booked themselves onto a bus tour through to Broome for a trip. Fred, their Son in Law, met them at Broome and helped with Lance as he was in a wheelchair for some of the time and the load was proving too much for Vera although she would not complain.

Shortly after their return from the Broome tour, Lance was admitted to the Gosnells Family Hospital. After a few days, he showed no improvement and was allowed to return home. By this time he was bedridden and needed constant attention, so with the help of the Silver Chain Nurses, Vera and her children took care of Lance in their home at Amaroo. The cancer that had caused the problem with his prostate had caused him, among other things, to lose the use of his legs. Whilst at home Lance's condition deteriorated quickly and he slipped into a coma. On the 23rd of September 1990, he quietly passed away in his home, with Vera alongside and also most of their family.

Clem and Ira were touring in their caravan with Hughie and Stella Manning and were able to return home in time to spend a few hours with Lance just prior to his passing. Lorna and Barry were returning from a caravan trip in the eastern states and arrived home a few days just before the funeral.

Lance was buried in the cemetery at Serpentine close to his parents' graves after a very moving Funeral Service at their Church in Armadale. During this service, Marjory and Neil read out the Eulogy, a copy of which is reprinted in this book.

In Lances will, he left most of the portions of his farming property to Neil to share with his mother and these, they would own jointly. The block of land at North Dandalup was left to be equally shared between Gwen, Marjory, Lorna and David. Neil continued to lease this from them and so still had full use of the entire farm.

Vera felt at a great loss as she had not only lost her husband and father to her children, she had also lost her best friend. After a period of immediate grief had passed, she tried to settle down to a new life but she just couldn't seem to be able to. She had many visits to doctors and specialist. She had been affected by a kidney and angina problem for years and now, these problems became worse. The medication to control the kidneys seemed to adversely affect the heart and vice versa. The doctors had trouble finding a suitable balance of drugs. On the 21st March 1991, she suffered some severe chest pains and had Lorna and Marjory take her to the emergency section of the hospital in Fremantle. While she was being attended to at the hospital, she passed away, in spite of the attempts of revival. She had lasted just barely six months after Lance had departed and now they could be together again, with God.

On 26th March, a funeral service was held for Vera in the Congregational Church in Armadale. At the conclusion of the service, the cortege continued to the Serpentine Cemetery where her body was interned into the same grave as Lance. Their children have erected a dual headstone to commemorate their final resting place.

Colin was booked into the hospital in Melbourne for a three month period for extensive tests to be carried out to determine the extent of the damage to his brain. The results from these tests allowed the medicos to formulate some operations that would help Colin. Although Clem and Colin travelled by plane to Melbourne, they travelled by coach when they spent some time with David (son of Spencer) Kentish at his property known as "Kalangadoo" while they were waiting between the exploratory sessions.

Clem returned to Melbourne with Colin a year later when they operated again.

Whilst the operation and treatment wasn't a total success, they were able to reduce some of the epileptic problems for Colin. Clem was with Colin for the whole time he was convalescing from the operation. Unexpectedly, Coralie flew over to be with Clem for a while. This would have been a big help to Clem to have someone else to share the burden of the situation with. After their return to the farm, Allan Uren began fortnightly sessions with Colin to help him relearn to read. Allan found that this helped Colin, not only in learning to read again but also in his outlook with other people.

The Head Injured Society of Western Australia has been very beneficial to Colin, with his attendance at the weekly sessions that he attended at Pinjarra. He also attended several of the camps which they hold from time to time and Colin fits in very well, as many of the other attendees are wheelchair bound and with him being a big strapping lad can offer a lot of assistance to these others. He is often referred to as the "legs".

There has been a lot of development in Mandurah, particularly in the vicinity of the grounds where the family had taken their annual holidays in the early years. Neil and Jacky along with their family frequently take their holidays in that same area but now they choose to stay at the Silver Sands Resort, which has been built close to the old campgrounds. After working physically and mentally hard for the year, it is a very pleasant break when you can stay in a place like this and totally relax and have no pressures.

Farming is growing a much faster business and with the constant necessity to keep abreast of the changes, one is required to spend much more time running the business than has been the case in the past. Good farmers of today need to be much more astute in their business dealings than in the past, just to remain afloat.

Frequently you hear the cry "the whinging farmer". But if those in other walks of life were to receive a return for their labours that was equal to that which they received 10-12 years ago but with today's costs, they would have ample reason to complain.

Diversifying on the land is one way to overcome this deficiency in income and Neil, along with many other farmers, has elected to use part of his holdings for the purpose of agroforestry. Prices that were presented at the time of planting for the finished crop indicated some good returns for the use of the land for this new purpose. By careful management within the timber industry and balancing the harvest between natural forest and agroforest, good returns can be expected. Once again the costs involved are at the outset of the operation and it would be at least 10 years before you would see any return from your input. This elapse of time also increases the risk factor, due to pest and fire, so it is still a big gamble.

An operation was performed on Ira's upper leg to remove a melanoma that had formed. Two years later the melanoma reappeared to the same area and was removed again. Further tests showed that there were some cancerous nodes present in the groin and later on, this condition had spread to infect her lungs. Although she was on medication, her condition deteriorated and she became bedridden. The family elected to have her at home so they could be with her and care for her together. Late in January 1994, Coralie moved into the adjoining bedroom so she could be nearby when her mother required attention. They had oxygen in bottles by her bed and it was given to her when necessary to assist in breathing. The sisters from the Silver Chain Hospice called in every 3-4 days to assist and advise where necessary.

The family had gathered in their home around Ira, when, on February the 18th 1994, she quietly passed away, ending her suffering.

The funeral service was held for Ira in the Chapel of the Fremantle Cemetery where a very large group of mourners attended to pay their last respects. Trevor Moffat (Esther's son in law) conducted the service and a eulogy was presented by Mrs Connie Senior following an address by Mr Peter Kargotich. Her mortal remains were cremated and sometime later, Clem spread these over the water of the ocean, in front of their beach shack at Mandurah, where Ira had spent many happy hours with family and friends.

It would appear that Colin has inherited some of his mother's gift. He has completed some works of art and these do show promise of better things to come.

On 28th of April 1994, Clem was requested to attend Government House in Perth. He was included in the Investiture Ceremonies that were to be held on that day. Her Majesty The Queen has been graciously pleased to approve him to be awarded the Medal Of The Order Of Australia. This medal was awarded to Clem to show appreciation for his past duties to the community and these were officially listed as ~

"President, Serpentine Jarrahdale Shire Council since 1964; Executive Member, Country Shire Council's Association of WA 1976Ä1984. Patron of many sporting clubs. Reformed and was President, Serpentine Cricket Club 1982. Visiting Justice, Karnet Prison Farm. Fire Control Officer since 1944 "

The farm cottage that was constructed in the late 1940s that is situated behind the main house on Bolinda Vale was refurbished during 1994. Deborah has the use of this house now and has made it her home. By being closer to her grandfather she is able to offer him some greater help.

During 1995 Colin made the move from living at home with his Dad to The Home of Peace annexure in Carlisle. Here he fits in with others who are in a similar situation as himself and does assist those who he can. He has time to develop his skills with art and this will be of great benefit to himself.

Neil and Jacky have a keen sense of community involvement and they are frequently out at night attending to meetings or playing sport. Neil has been president of the Serpentine Badminton Club for about 12 years and received a Life

Membership in 1982. He also has been the president of the Serpentine Tennis club for many years with Jacky being the secretary for some of those years. They still play at the Serpentine Tennis Courts although at present the courts are undergoing a concrete resurfacing. Jacky continues to play badminton with the Serpentine club during the winter but Neil plays badminton only a few times each year.

Clem is still working with Coralie and Graham on Bolinda Vale and their adjoining properties. They concentrate their time to the production of quality beef cattle which are bred from a very good stock of the Angus, Hereford and Limousin breeds. The hand feeding of these, particularly during the summer months, is the most consistent job. Most of the previous heavy work is now performed by machinery but days are still very full. Colin was able to offer some help with some of the work until he moved to Carlisle, and Deborah, Coralie and Graham's daughter, is working with them. As is Michael, now that he has completed his education at Harvey Agricultural School at the end of 1995.

Although they still have a full complement of farm equipment to carry out any of the jobs required from time to time Clem has opted to use contractors for super spreading and some of the larger jobs.

They enjoy their time working together and this is shown at hay time when they would have 3 hay balers working at one time. Clem at 80 still drives the square baler at hay time and he, Coralie and his Granddaughter, Deborah, frequently have races to see who can bale the most hay over a given time. This not only improves production efficiency it also gives them a sense of friendly competition and companionship.

When it comes to feeding out of the hay they have implements attached to the tractors to handle several hay rolls at one time mechanically. Clem puts a few small bales in the back of his ute and works with the others by dishing out small amounts of hay to those cattle separate from the others, but most times he just sits on a bale and watches the others work.

Stock handling work takes up a good part of their time and Coralie's neighbour and friend, Bev, is a great help with this work. She helps to make up a team of good workers that are needed to effectively run this farming property.

Clem and Coralie have added to their holdings over the years which is now in the vicinity of 4,000 acres. With several hundred head of cattle, you can appreciate the time and effort involved to run an operation of this size.

The care of the land, both agricultural and other, is of great concern to the community. Neil is involved in the Serpentine-Jarrahdale Land Conservation District Committee. For a period of 6 years, he held the position of secretary-treasurer. At the present time, he holds the position of treasurer. The group is working hard to make people aware of the land's capabilities and trying to revegetate a lot of open and degraded areas within the shire.

The Serpentine Land Conservation District Committee is one of 120 throughout the state. These have been gazetted by the government and have a range of options and powers at their disposal in the area of conservation and care of the land. By careful planning and preparation, the Land Conservation District Committees are able to attract considerable government funds to carry out some very important work in this area.

Stray dogs are still posing a problem with livestock. Recently, while Coralie and Clem were away on holidays, Graham and Michael had a problem with dogs. They noticed some dead sheep in a paddock near the hills and on closer inspection saw a pair of Rottweilers mauling and killing a number of sheep. The dogs were shot and through the local ranger, the owners were found. Compensation was paid by the owners of the dogs. They had lost control of the dogs which had been missing for several days.

Josephine has completed her primary school education and next year she will attend the college of Frederick Irwin in Mandurah for her secondary education. Stuart will complete his final year of primary school at Serpentine in 1996 and then he also will attend Frederick Irwin and in 1997 will commence his secondary education.

Graham has a flair for things mechanical and spends a lot of his spare time to the restoration of machinery. He has successfully restored an early model Fordson tractor, shearing plant, diesel and petrol engines and recently has just completed a working model of a J Fowler steam traction engine. This you will see at shows and fairs towing a model fairground organ that he has recently acquired.

1996 and the future?

The Kentish's of Keysbrook will remain in agriculture for a good many years yet but the continued increase of costs will make things increasingly difficult. Continued diversification will help to offset this.

April of 1996 will be the 70th anniversary of the Kentish's arrival in Keysbrook.

In the near future, we will see the Serpentine Jarrahdale Shire increase from the current 9,000 inhabitants to an expected 60,000. This will impose a greater pressure on those involved in agriculture in the district.

We have another new generation coming along now and they will learn improved practices, both in agriculture and other avenues of life, to carry on into the future.

Being aware of our history is important as we can record and recognise our mistakes and make certain that they do not reoccur. Our history can tell us that which has been done well and can indicate areas where improvements can be made for a better, stronger and brighter future.

I am hoping that what I have been fortunate enough to be able to record here is of benefit to those in the future as a record of our past. Not only in the recording of Kentish's of Keysbrook family details but also in showing where life and farming practices have altered over the period of time since their arrival in the district.

I offer you a salute to the future and wish everybody to enjoy the best of health, wealth and happiness!

Western Australian Property Details.

"BOLINDA VALE" Approximately 60 Km south of Perth Purchased March of 1926. This was approximately 640 acres of which about 50 acres were cleared and suitable for cropping and grazing. Some of this property formed the large hill to the east and was not cleared. At the time of dissolving the partnership, 100 acres of this hill was vested to Lance and Vera.

"EMMITTS" This property was purchased about 1931 - 1932. It was purchased by Clem and Lance. About 450 acres most of which was bush with some fencing. This cost about, £3/-/- per acre for the cleared portion and, £2/10/- for the uncleared section.

* "NORTH DANDALUP" Cockburn Sound Loc. 16 of 93 acres was purchased from a Mrs Dunlop in 1935 and is on the corner of the railway reserve and Readhead Road 2 miles north of North Dandalup and runs south along the eastern edge of the railway line to the North Dandalup River.

* "MOUNT VIEW" Murray Loc. 249 Approximately 1.5 Km north of Bolinda Vale. This was purchased in 1944-1945 from Mr Wells (He purchased from Mr Opitz).

"GOBBY'S" adjoined Bolinda Vale to the south. This was about 160 acres and cost, £6000/-/- or, £1/16 / 10 per acre in 1947-1948. This was the property that was owned by Mr Mathews as mentioned previously and he sold to Mr Gobby who was a "St. George's Terrace farmer ".

* "FISHERS" block of 160 acres was purchased in about 1950. This block was owned by Fred Fisher and the adjoining block

(across Fisher Rd.) was owned by his brother Maurie who sold to Mr Norm Fletcher.

* "HECTORS" This 200-acre block is at the end of Fisher road and was purchased about 1952.

"The BLOCK" This is on Hopelands Rd. and originally consisted of three square miles (1920 acres). This was purchased in about 1953.

* "THE NEW FARM" Peel Loc. 824 of 72 acres was the first land that Lance and Vera purchased after the Partnership of Kentish Bros. was dissolved. The titles show that this was purchased on the 8th of December 1961.

* "ARNDTS" was purchased in 1966. This was divided by Utley Road with approximately 100 acres on the south side and 200 acres on the north side of the road.

"Bees" Purchased from Mr Tom Bee. This extends from the creek at Keysbrook, north to Emmitt's and includes that area surveyed for the township of Keysbrook.

"Pollards" Across the railway line from Emmitt's and continues to Elliot Road, west of the Keysbrook railway station.

"Le Bar" Approximately 320 acres 3 kilometres south of the main farm. This was a portion of the original Ingpen Estate. Purchased from Mrs Le Bar in 1985.

"Hill" A small hill block became available adjacent to Kingsbury Drive. Clem purchased this and set up and occasional use Caravan Park on it. Caravan clubs and family have the use of this from time to time.

65 acres alongside and east of the highway, adjoining the north boundary of Bolinda Vale was purchased in 1993.

(* These properties are owned by Lance and Vera, either being transferred at the time of the dissolving of the partnership of KENTISH BROS.. or purchased after that time. The other properties were to remain with Clem and Ira.)

Western Australian Map

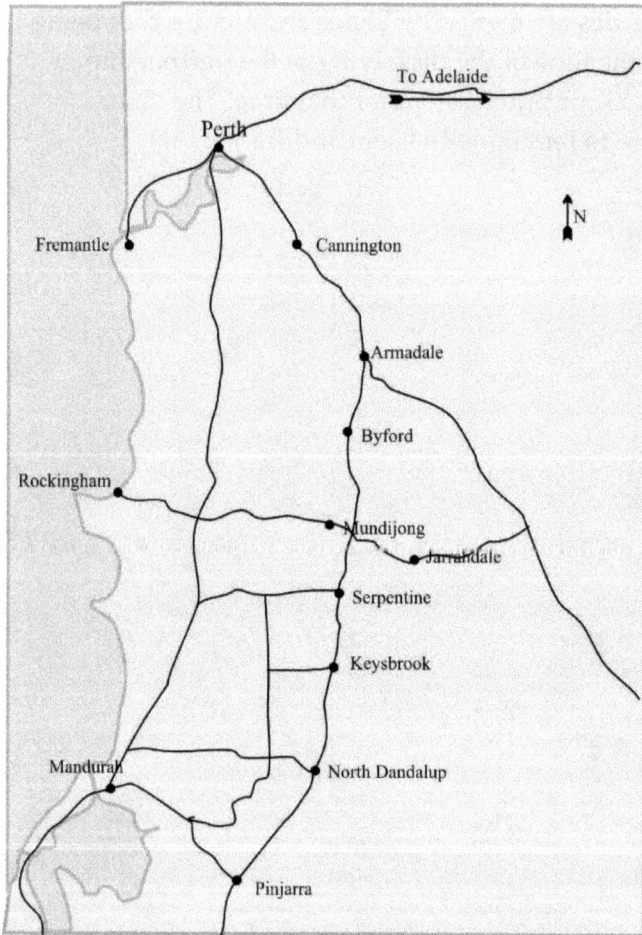

Western Australia

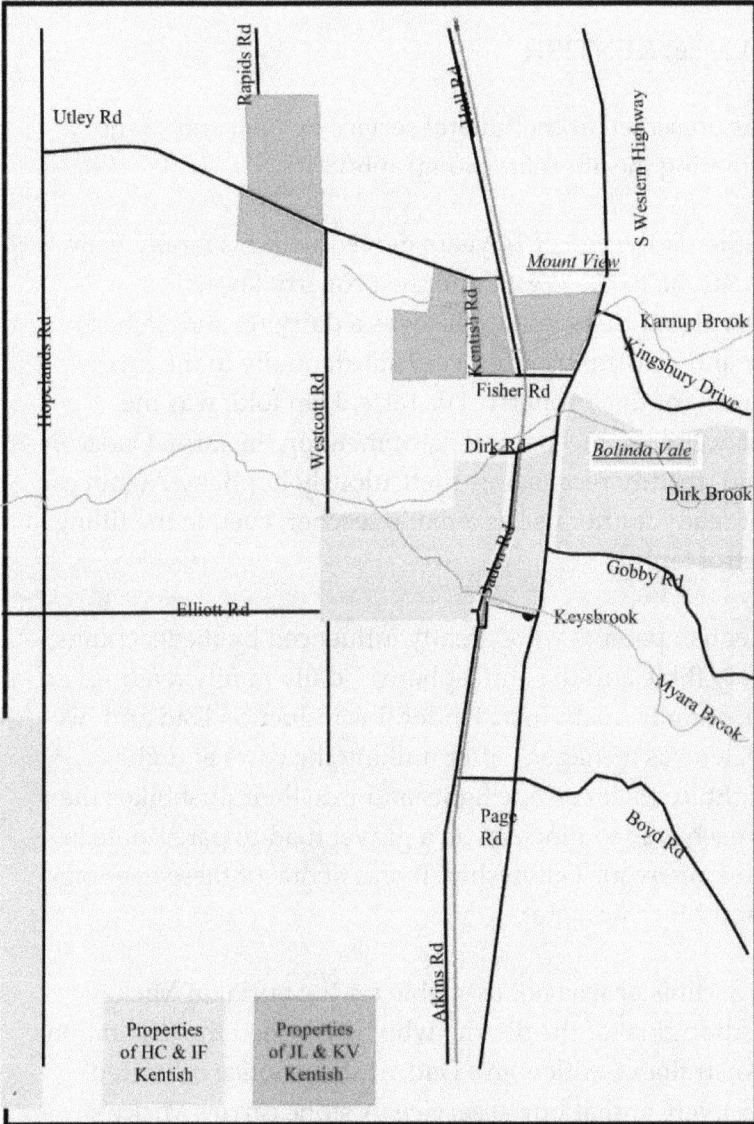

Western Australian Properties

Eulogy
Mr J L (Lance) KENTISH

(This was presented at the funeral service by Neil and Marjory, each delivering the alternate paragraph.)

1. My Father at the age of 14 years moved with his family from South Australia to the West, onto the property known as "Bolinda Vale" at Keysbrook. This was a dairying and orchard property and Dad contributed very substantially to the further development of this property. His forte, I am told, was the maintenance of the mechanical equipment on the farm. Dad's sister Enid, readily recalls how meticulously her Baby Austin car was kept ready for her use as a piano teacher, even to the filling of the petrol tank.

2. The Kentish parents were greatly influenced by the teachings of the Holy Bible and the atmosphere of daily family readings as part of the family's activities. Under this influence, Dad and his brother Clem, as teenagers, after milking the cows at night, would light up their carbide lights and ride their pushbikes the 17 miles each way to Pinjarra, on a gravel road to participate in Christian Endeavour Fellowship. It was at one of these evenings that he and Mum first met.

3. When a Minister was not available for the burial of Mickey Ball, an Aboriginal of the district who had earned the reputation of "An Australian Gentleman", Dad, at short notice officiated very effectively at that burial service. A stone carries Micky's epitaph in the Serpentine Cemetery, placed there by Mr Drake, a member of the community.

4. Dad eventually became Sunday School Superintendent at Serpentine, later sharing this with Mr Laurie Manning and then commencing the Sunday School at Mundijong.

5. Dad was also Lay Preacher for the Pinjarra and Jarrahdale circuits of the Methodist Church and I can remember travelling on several occasions with the family when Dad was taking the service at Waroona, Mandurah and Mundijong. The Keysbrook Hall was also used as a Church but this hall was demolished some years ago.

6. Dad and his brother, Clem built up a large dairy herd in their partnership on the Bolinda Vale Farm, this being the largest whole milk contract for a number of years. Because of their good management and ability to work, they were able to expand and capitalise on the opportunity. When the first milking machine was installed in the early 1940s, Granddad Kentish said: "You boys are getting too modern". Before this, the herd of 40 odd cows was milked by hand twice a day, I'm glad I wasn't around to do that, it would have been hard work and Aunty Esther was the best at hand milking.

7. Dad did not smoke or drink, although I do remember when Masters Dairy used to bring out their Christmas gifts, there was an offer of 2 boxes of chocolates or 2 bottles of sparkling wine. We kids only ever saw the chocolates.

8. In the early 1960's Dad bought a small boat, a 14 footer if I remember correctly, fitted with an 18 horsepower outboard motor and we kids all learnt to water ski. We used to go to Palm Beach mostly, occasionally to Deep Water Point. Later we updated to a bigger boat, this time a 16 footer with a 40 horsepower outboard motor, which we thought was marvellous as now we could ski much faster. We used to ski from Palm

Beach to Kwinana and sometimes to Garden Island and return. A lot of this sport was done after the evening milking and some between milkings on the weekend.

9. Dad became involved with the Methodist Boys Farm at Werribee, near Wundowie and because of his practical farming and sound business knowledge he made a very useful input into this project. When this farm was sold, his time and efforts were put into another Children's project, The Mofflyn Children's Home in Victoria Park. He also gave a substantial donation to assist in the further education of the children.

10. For 18 years our dad was on the Executive Committee of the Farmers Union and for 8 of these years was president of the Whole Milk Section. This was a very time-consuming activity but the farm work did not suffer because of this.

11. In 1955 a new house was built on Mount View and this was the start of Uncle Clem and Dad dissolving their partnership. After a few years, a dairy was built and the dairy herd halved and the two farms managed independently. We then commenced to build up an A.I.S. herd of which Dad was very proud. In the mid-1970s the milk quota was sold along with some of the dairy cows and a beef breeding program and the herd was commenced. All in all, Dad milked cows for 49 years.

12. In 1980 Mum and Dad had a house built in Armadale and moved there from the farm. Dad continued to take an active interest in the farm work. It was about this time that the more extensive caravan trips were taken. Poor old Neil was left to "batch", again !!!

13. In 1965 my Dad was appointed as Justice Of The Peace and for many years was involved as a Visiting Justice for the Karnett

Prison Farm. His duties as a Justice Of The Peace continued until just a short time ago.

14. Mum and Dad bought a caravan in the 1960s and I suppose got the "Travel Bug" and that developed into some sort of a ritual. They would take off so often that they only seemed to come home to mow the lawn. They have travelled Australia extensively in the last 20 years or so with their car and caravan and have made many friends in the caravanning clubs. I have heard it said that they were lucky, I suppose they did have some luck, but they had worked very hard on the farm over the years and I consider that they earned these trips.

15. Possible the most pleasant, interesting and satisfying task he undertook was being Father Christmas for the children of the Keysbrook and Mardella at their Christmas Tree celebrations.

16. Over the years our Dad was always ready to help his neighbours who had the misfortune, especially with fires and so on and his help during the Dwellingup fire was extensive, along with the donation and delivery of hay for stock-feed after this disaster.

17. As Dad's health deteriorated Mum and Dad bought a unit in the Amaroo Retirement Village, moving there only in June of this year. Dad had his last holiday, which he dearly wanted to do, only 6 weeks ago when they flew to Broome and then travelled by coach to Carnarvon and then back to Perth.

18. His last wish was to be at home with his family and not stay in the hospital. With the great help of the Silver Chain Hospice people, we as a family were able to manage this and to achieve his last wish "to die at home".

He passed away peacefully last Sunday the 23rd of September 1990

Photos of Importance

Mr. Edward James Kentish 1895	Miss Edith Alice Nicholls 1905

The Kentish family. *L-R:* Enid, Edith, Lance, Esther, Ted, Clem

| Miss Enid Winifred Kentish 1931 | Mr. James Lancelot Kentish 1933 |
| Mr. Herbert Clement Kentish 1935 | Miss Esther Bell Kentish 1938 |

Lance and Vera's wedding group. *L-R:* Clem, Esther, Lance, Vera, Connie, Keith

Esther and Allan's wedding group. *L-R:* Cooper, Gwen, Allan, Esther, Enid

Enid and Reg's wedding group. *L-R:* Clem, Dorothy, Gwen,
Reg, Enid, Marjory, Esther

Clem and Ira's wedding group. *L-R.* George, Meline, Clem, Ira, Valma, Baden

Lance and Vera and family at their new home 1955

Clem and Ira with their family

Reg and Enid and family

Esther and Allan with their family

Acknowledgments

Mr Peter Kentish   Passage for Prosperity
Mrs Gwen Swaby   Compilation of Family Tree
Mrs Enid Ingpen   Grandmother Remembers
The Bell Family Tree Committee   Bell Family
Pioneers Assoc. of S.A.   Stanley Grammar School
Mr Eric Senior   Notes on Clem's Shire activities
Mr Hugh Manning   Interview

Other publications to read: ~
"The Serpentine" Mr Neil Coy
Contact.........................The Shire of Serpentine Jarrahdale,
Mundijong Western Australia 6202
"Plains of Contrast" Committee
Contact.........................Mrs. G Foulis, PMB 21, Orroroo South
Australia 5431
"The Bell Family 1728 - 1986" Committee
Contact.........................Mr. Ronald C Bell, 38 Strathcona Ave.,
Clapham South Australia 5062
"The Joys and Sorrows of a Migrant Family" Mrs Florence
Edwards
Contact.........................
"Passage for Prosperity" Mr Peter Kentish
Contact.......................Mr. Peter Kentish, Harvey St. East,
Woodville Park, South Australia 5011

N.B. *Unfortunately in 2020, some of these may be unavailable.*

David Kentish was brought up on a dairy farm just south of Perth in Western Australia near the small settlement of Keysbrook.

Before the time of broadcast television, his father, J. Lance Kentish, spent time in the evenings inventing and telling stories about the bush animals, the talking red-gum tree and the magic carpet to his family.

David has continued in this same vein with the telling of stories of imaginary Australian bush animals and friends.

He has completed two books of family history, **The Kentish's of Keysbrook** and, *The Kentish's of Keysbrook- The Next Generation.* These two books he has self-published.

David, with his wife Barbara, enjoys travelling with their caravan in and around the Australian outback and bush. This is where he gets most of his inspiration which has led to a collection of stories.

David has also completed an Australian adventure book, **King's Gold** or *Dramatic interruptions to a well-*

planned adventure, which takes the reader through the outback and into exciting situations.

Another booklet, **Beside the Billabong** or *The Adventures of Warragul the Bunyip* is a story of Warragul, who is a juvenile bunyip of Australian mythology and his animal friends who find the fun and adventures of living at a billabong.

A Place Called Earth or *The Amazing Booklet of David's Enlightening Short Stories* was written whilst on one of these trips and is an exciting story of how the earth began and developed over many countless years. A well-written story that will keep you intrigued right up to the last page.

His mother, Vera was a big impact on David's life and he has completed **KVK** *A Life Well Lived,* a biographical look at her eventful life.

David has several other works in the pipeline so keep a lookout for more stories by David Kentish.

For more details contact the author dkentish@westnet.com.au or visit his website at www.davidkentish.com.au

www.ingramcontent.com/pod-product-compliance
Lightning Source LLC
Chambersburg PA
CBHW052005090426
42741CB00008B/1558